Cyber Physical Human Systems

Scrivener Publishing
100 Cummings Center, Suite 541J
Beverly, MA 01915-6106

Publishers at Scrivener
Martin Scrivener (martin@scrivenerpublishing.com)
Phillip Carmical (pcarmical@scrivenerpublishing.com)

Cyber Physical Human Systems

Emerging Trends and Applications

Edited by

Anitha Velu

Prasanth Aruchamy

Raghu Ramamoorthy

Rajesh Kumar Dhanaraj

and

Manjula Arunraj

Scrivener Publishing

WILEY

This edition first published 2026 by John Wiley & Sons, Inc., 111 River Street, Hoboken, NJ 07030, USA and Scrivener Publishing LLC, 100 Cummings Center, Suite 541J, Beverly, MA 01915, USA

For more information about Scrivener publications please visit www.scrivenerpublishing.com.

Wiley Global Headquarters
111 River Street, Hoboken, NJ 07030, USA

For details of our global editorial offices, customer services, and more information about Wiley products visit us at www.wiley.com.

The manufacturer's authorized representative according to the EU General Product Safety Regulation is Wiley-VCH GmbH, Boschstr. 12, 69469 Weinheim, Germany, e-mail: Product_Safety@wiley.com.

Library of Congress Cataloging-in-Publication Data

ISBN 978-1-394-40288-5

Cover image: Generated with AI using Adobe Firefly
Cover design by Russell Richardson

Set in size of 11pt and Minion Pro by Manila Typesetting Company, Makati, Philippines

Printed in the USA

10 9 8 7 6 5 4 3 2 1

Contents

Preface

Cyber Physical Human Systems (CPHS) has emerged as one of the most promising multidisciplinary fields of the modern era as a result of the quick development of intelligent technology, which has changed how people interact with machines and the physical world. In order to give readers a thorough grasp of the ideas, architectures, technologies, applications, and societal ramifications related to CPHS, the book *Cyber Physical Human Systems: Emerging Trends and Applications* has been meticulously crafted. This book demonstrates how intelligent systems combine sensing, communication, computation, automation, and human interaction to create intelligent, adaptive, and secure settings by bringing together contributions from many research viewpoints. Human–machine interfaces, supervisory control systems, deep neural network-based learning models, digital twin architectures, Industry 5.0 applications, and intelligent healthcare systems are among the advanced topics that the book progressively moves toward after starting with the fundamental overview and fundamental principles of CPHS. The information is useful for academic researchers, engineers, industrial practitioners, and students alike because each chapter is organised to offer both theoretical comprehension and practical application.

This book's transdisciplinary and application-focused approach is one of its main advantages. Together, the chapters highlight how AI, IoT, edge and mist computing, secure communication systems, robotics, autonomous cars, healthcare technologies, and cyber-physical security frameworks are integrated. Emerging technologies that are anticipated to influence the next generation of intelligent systems, like 6G-enabled autonomous vehicle networks, flexible sensor-based smart health systems, and digital twin-driven cyber physical architectures, have received particular attention. This book discusses important issues pertaining to data privacy, ethical issues, legal ramifications, and social effects of CPHS deployment in practical settings. By relating research ideas to actual case studies, intrusion detection systems, and cooperative human-robot workplace models, the book is further strengthened. Readers can comprehend both the development of intelligent cyber physical systems and their responsible deployment for safe and sustainable technological advancement thanks to this fair coverage.

The goal of this book is to be a valuable resource for future study in the field of intelligent CPHS as well as for fostering creativity and multidisciplinary cooperation. By enabling safer transportation systems, effective industrial automation, intelligent healthcare monitoring, secure communication networks, and human-centered smart environments, the technologies covered throughout the chapters offer enormous potential to improve people's quality of life. The book helps academicians and practitioners to create technologically sophisticated, morally sound, and socially advantageous solutions by addressing both possibilities and challenges. For the benefit of society and future generations, it is anticipated that this compilation will encourage readers to investigate novel avenues in CPHS research and help create intelligent, robust, and sustainable systems.

1

Cyber-Physical Human Systems: Overview, Fundamental Concepts and Perspectives

Anitha Velu[1]*, Raghu Ramamoorthy[2], R. China Appala Naidu[3] and Rashad G. Abaszade[4,5]

[1]*Department of Electronics and Communication Engineering, Sri Sairam College of Engineering, Bengaluru, Karnataka, India*
[2]*Department of Computer Science and Engineering, The Oxford College of Engineering, Bengaluru, Karnataka, India*
[3]*Department of Computer Science and Engineering, M S Ramaiah Institute of Technology, Bengaluru, Karnataka, India*
[4]*Department of Electronics and Automation, Azerbaijan State Oil and Industry University, Baku, Azerbaijan*
[5]*Ecology Department, Azerbaijan University of Architecture and Construction, Baku, Azerbaijan*

Abstract

A new field of study called cyber-physical human systems (CPHS) examines how physical elements, humans, and cyber technologies interact intricately to accomplish particular goals. From a technological standpoint, CPHS can both uncover technological concepts and build the technological infrastructure for intelligent manufacturing. A cyber-physical system (CPS) facilitates smooth communication between virtual and physical worlds by fusing digital technology with tangible activity. In a variety of verticals, including manufacturing and oil and gas, this sector improves sustainability, security, and productivity. Because of composite relations among human, physical, and cyber elements, the development and implementation of CPHS raise important ethical, legal, and social issues (ELSI). This chapter presents a brief run-through of the fundamental concepts, characteristics, and perspectives of CPHS. Additionally, this chapter also gives a detailed description of all the topics covered in the book, including human-in-loop, role of CPHS in digital twin, Internet of Things (IoT), deep neural networks (DNNs),

**Corresponding author*: aniveluece@gmail.com; ORCID: 0000-0003-2683-4495

Anitha Velu, Prasanth Aruchamy, Raghu Ramamoorthy, Rajesh Kumar Dhanaraj and Manjula Arunraj (eds.) Cyber Physical Human Systems: Emerging Trends and Applications, (1–20) © 2026 Scrivener Publishing LLC

6G communication, mist, and edge computing by addressing the problems in data privacy and security for CPHS for real-time implementations.

***Keywords*:** Human-machine interface, cyber-physical human systems, computation and communication, user interface, trust and ethics

1.1 A Run-Through about Cyber-Physical Human Systems

Cyber-physical human systems are those that integrate physical elements, human knowledge, and cyber technologies working together to achieve a common objective. Humans are seen as essential components of complex cyber-physical systems, which incorporate networking, computation, communication, and control technologies into the underlying physical system [1]. This is in contrast to the commonly held belief that people are autonomous, passive beings that use, consume, or run these systems. The significant advancements in computers, automation, communications, and control laid the groundwork for the field of CPHS [2]. By examining methods to integrate important cognitive processes like perception, attention, and memory into the CPHS framework using concepts from cognitive neuroscience, recent research findings are further pushing the envelope.

These systems, combining human interactions, computational intelligence, and physical processes, are transformational, but they also raise issues that need to be resolved to guarantee their fair and responsible usage, which is now a key concern [3]. Numerous significant and vital sectors have been dominated by CPHS in recent years. For example, CPHS is used in manufacturing and production to encourage automation and accuracy. The healthcare industry also uses them in state-of-the-art medical devices [4]. The plan and development of CPHS in many settings, at various levels, is the fundamental component of intelligent manufacturing. As information technology advances, intelligent manufacturing is evolving into New-Generation Intelligent Manufacturing (NGIM).

The prevalence of CPHS is highly important because, in recent years, it has expanded across all sectors and industries. Since increased connectivity often leads to a more serious attack, it is more important to concentrate on its impacts and solutions to protect and preserve them. The architecture of CPHS is illustrated in Figure 1.1, where the human and physical world can create a bridging gap between human and machine [5]. While CPS serves as an interface between cyber systems and physical systems, human cyber systems (HCS) serves as a link between human knowledge and cyber systems. Here, physical and human systems are interconnected through human physical systems (HPS). One benefit of CPHS is that it replaces a

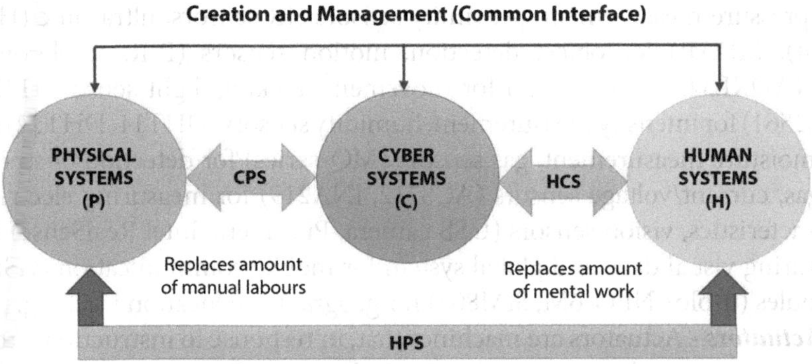

Figure 1.1 Architecture of the CPHS system.

significant portion of human mental strength and reduces manual effort, which improves the human-machine interface (HMI).

The architecture of cyber-physical systems combines software and hardware elements. This architecture makes it feasible for physical processes and computational components to interact dynamically. This serves as the cornerstone for CPHS's perceptive and accommodating actions. The following sections describe the components of a CPHS architecture:

1.1.1 Physical Process

The physical process of CPHS is a real-world physical phenomenon or system that the cyber (computational) components are monitoring, controlling, or interacting with, i.e., the environment that CPHS interacts with in the real world [6]. In general, a physical process is any component of the system that includes physical dynamics that may be sensed, controlled, or acted upon by digital systems, such as motion, temperature, pressure, or any other observable physical behavior.

1.1.2 Sensors and Actuators

Equipment that gathers information from physical processes, such as temperature, pressure, and motion, is known as a sensor [7]. Sensors and actuators are primary interfaces among substantial and computational components of CPHS. Sensors gather information from the physical world, whereas actuators respond to decisions made by the cyber system by modifying the physical environment. Some of the most commonly used sensors and actuators involved in the development of CPHS are listed below:

i) Sensors - Temperature sensors (DHT11, LM35, thermocouples) for heat/thermal measurements, pressure sensors (BMP180, MPX5010) for fluid/

gas pressure measurement, proximity sensors (IR sensors, ultrasonic (HC-SR04), LIDAR) for object detection, motion sensors (PIR, accelerometers (ADXL345), gyroscopes) for movement tracking, light sensors (LDR, TSL2561) for intensity measurement, humidity sensors (DHT11, DHT22) for air moisture measurement, gas sensors (MQ-series) for detection of smoke or gas, current/voltage sensors (ACS712, INA219) for measuring electrical characteristics, vision sensors (USB camera, Pi camera, Intel RealSense) for capturing visual data, and global system for mobile communication (GSM) modules (u-blox NEO-6M, SIM808) for geographical location tracking.

ii) Actuators - Actuators are machines that, in response to instructions from the computing nodes, regulate the physical process. Some of the commonly used actuators in CPHS are electric motors like direct current (DC), stepper, and servo motors that help to produce any rotary or linear motions; relays such as mechanical relay and solid-state relay are used for switching purposes; solenoids like door locks and valves convert electrical signals to linear motion; Peltier devices and heating coils are used to implement heaters/coolers; light-emitting diodes (LEDs), smart bulbs, and lamps are used as indicating devices or illumination adjustments; organic light-emitting diodes (OLEDs), 7-segment displays, and liquid crystal displays (LCDs) are used to present information to the user; speakers/buzzers are used to provide alerts; and solenoid/moisturized valves are used to control the flow of fluid/gas.

1.1.3 Data Transmission

Data transmission, which helps in transferring data between the physical and virtual worlds, is accomplished by a communication network, which can be wired or wireless. Point-to-point channels and data flow play a vital role with respect to this transmission [8]. These channels are today far more likely to be a component of a wireless network than they were in the past, when they were implemented using copper lines. Both analogue and digital data transmission are used by CPHS, depending on the part, application, and processing step. However, because of its accuracy, dependability, and simplicity in integrating with computers, digital transmission predominates in the majority of contemporary systems.

The amount of data transferred in a given amount of time is known as the data transfer rate. It establishes whether a network may be used for tasks requiring complex, data-intensive applications. Inadequate infrastructure, server health, latency, and network congestion can all affect overall performance and reduce data transmission rates to below-average levels. High-speed data transmission rates are required to process complex operations like large file transfers and online streaming. The role of analog transmission is to capture real-world signals before conversion, used in the physical layer or initial sensing/actuation, whereas digital transmission helps in reliable communication between the system components used in communication, processing, and storage.

1.1.4 Computational Nodes

In a distributed CPS or CPHS architecture, a computational node is a processing element (such as a microcontroller, embedded system, or cloud server) that processes sensor input, runs algorithms, and interacts with other parts [9]. It is an essential component of a distributed system that pools resources to enable intricate calculations, such as scientific simulations. In CPHS, a computational node is a virtualized computing resource, which can be a component of grids or clusters, that users can access whenever they need to. Equipment that processes and analyzes sensor data, such as servers and microprocessors, is also referred to as a computational node. Computational nodes take care of data acquisition, pre-processing, analysis decision-making, control actions, communication and coordination, human interaction interface, system monitoring, and diagnostics.

1.1.5 Control Algorithms

In CPHS, a control algorithm is a mathematical model or software-implemented logic that chooses the right actuator actions based on sensor data in order to maintain system stability or produce a desired behavior. Software that evaluates sensor data, makes judgments, and produces control signals is known as a control algorithm. Since computers were created, software has swiftly taken over our daily lives. In reality, practically all contemporary engineering systems integrate cyber and physical systems [10]. These technologies integrate cyber components into our physical world. The combination of physical and digital elements presents both new possibilities and difficulties. This kind of integration creates a new point of service in conventional physical items like wearable technology, biochemical process intelligent control systems, and automobile brakes and engines.

However, the incorporation of cyber components introduces additional levels of complexity, which could significantly complicate their design in order to ensure performance. The hybrid nature of CPHS consists of both discrete and continuous mechanisms since physical and a package (software) are integrated together, which should be acknowledged by a suitable mathematical model. Hybrid dynamical systems use state machines in order to simulate cyber machineries and dynamical compositions for carnal representation. Hybrid-type dynamical representations can generate precise forecasts and facilitate evaluations of CPHS's accomplishments.

The phases of digital manufacturing and digital-networked manufacturing have already been completed. In order to assist developers, academicians, and researchers in creating diverse algorithms and designs for a wide variety of applications through the deployment of CPHS systems, this book attempts to address recent technological breakthroughs and applications in the field of intelligent CPHS. Social connections, workplaces, and everyday lives are all changing as a result of CPHS. These intricate systems that combine hardware, software, and networks are at the heart of numerous crucial industrial applications, from manufacturing and industry to healthcare and government. They enhance the interconnectivity of gadgets and systems across sectors to increase productivity and maximize effectiveness.

1.2 Fundamental Concepts of CPHS

Cyber-physical human elements are dispersed and intricately linked, which include components for detecting, controlling, processing, calculating, communicating, and actuating. CPHS are safe, interoperable, smart networked systems interconnecting both human users and the physical world in real-time. CPHS can be modest and resident, or it can be massive, complicated, and highly networked by spanning multiple domains. Figure 1.2 describes the workflow of CPHS, where the intelligent manufacturing mode with human-cyber-physical collaboration happens for any intelligent manufacturing. The integration of computation, communication, physical processes,

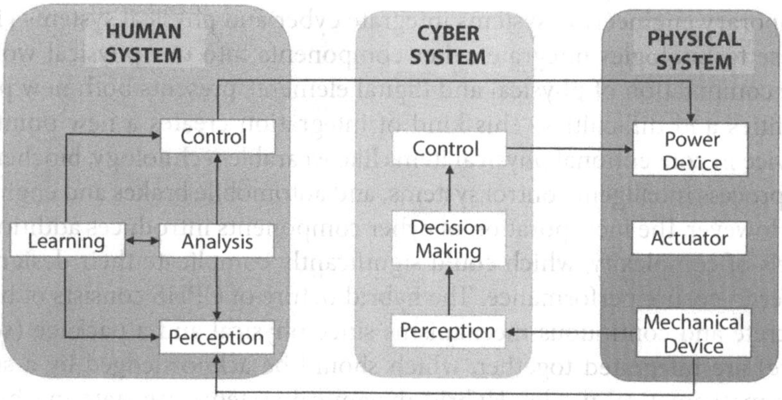

Figure 1.2 Interaction of CPHS with cyber, physical, and human systems.

and human interaction is one of the core ideas of CPHS. By proactively integrating humans as essential elements, whether as users, operators, or partners, CPHS expands upon CPS. The core concepts of CPHS are listed below:

1.2.1 Cyber-Physical Integration

The fundamental idea behind CPS and CPHS is cyber-physical integration (CPI), in which physical processes and computing (cyber) components are closely intertwined. It describes how a system's physical domain (hardware, sensors, actuators, and environment) and cyber domain (software, algorithms, and communication) work together seamlessly. By establishing a close, real-time interaction loop involving sensing, computing, communication, and actuation, cyber-physical integration is accomplished. This is made possible by network technologies, hardware, and software that let cyber systems (like control algorithms) observe and affect physical systems (like machines or the working environment).

1.2.2 Human-in-the-Loop

A collaborative approach called human-in-the-loop (HITL) integrates human knowledge into the creation of machine learning (ML) and artificial intelligence (AI) systems. Humans actively provide helpful guidance, feedback, and annotations to ML models during training, testing, or operation [11]. HITL aims to increase the accuracy, reliability, and adaptability of ML systems by leveraging the unique capabilities of both humans and machines. Even though ML models are incredibly powerful, they can nevertheless be enhanced by human knowledge in situations that call for judgment, contextual awareness, and the handling of insufficient data. By adding human input and feedback to the ML process, HITL fills this gap. Models can adjust to changing user preferences and real-world situations due to this human collaboration. ML systems are enabled to handle the refinement and complexity that are frequently ignored by strictly computational methods by incorporating the human factor.

1.2.3 Sensing and Perception

CPHS integrates technologies into architectures that allow the physical and computational domains to interact in real time. The perception layer is made up of sensors and actuators that gather information and carry out orders to operate as an interface with the physical world [12]. The control layer is made up of computational components such as real-time processors and embedded systems. Through specialized networking protocols, the communication infrastructure makes it possible for sensors, controllers, and actuators to reliably exchange data. When combined, these elements enable minimal human involvement in the monitoring, analysis, and manipulation of physical environments.

1.2.4 Computation and Control

CPHS generally refers to the networked systems where the computational component (cyber) is closely integrated with the physical component. In other words, the system and environment are sensed by the computational components, which then continuously offer feedback to regulate the system and act on the environment. New capabilities are produced by the interaction of physical and computational systems. Since it offers the back-plane that ensures prompt information transfer from the physical to the computational world, CPHS is widely preferred for computation and control. The network is said to be an essential part of CPHS and helps in the transmission of commands from the computation system to the physical component.

1.2.5 Communication and Networking

Wireless sensor networks (WSNs) and control models are natural extensions of the CPHS paradigm. It makes it possible for the computing environment to effectively monitor and control the physical systems [13]. Only effective networking and communication between the real and virtual worlds can support such an interface and enable seamless interactions.

The primary components of communication in CPHS are cyber-to-cyber, physical-to-cyber, human-to-cyber, cyber-to-physical, and cyber-to-human. These modes help with exchanging data or information, coordination, data collection, control, and decision-making. In CPHS, networking refers to the systems and procedures that make communication possible in a safe and effective manner, which include IoT, HMI, sensor networks, and wired and wireless networks. Networking makes it possible for the components to be connected to one another and guarantees dependable, real-time data flow.

1.2.6 Safety, Security, and Privacy

CPHS are utilized in a diverse field to facilitate previously unattainable convenience and progress upgradation. The combination of analog physical processes and networked digital systems creates unique characteristics that change how security theory is applied [14]. The crossover reaction presents unpredictable weaknesses; CPHS cannot be protected by cyber-security or physical-security ideas alone. While cyberattacks can result in physical flaws that have real-world consequences, physical attacks can also grant access to valuable assets and cause harm or information loss to the device. Security and privacy are therefore significant factors in planning, creating, and functioning CPH systems.

1.2.7 Adaptivity and Learning

Humans and cyber-physical components work together to complete tasks in adaptive CPHS, and they adjust to changing circumstances to achieve mission objectives. The ML model integrated with CPHS can self-evolve with copious data in a self-labelling fashion during deployment, which is one of its noteworthy features [15]. Catastrophe of human-machine interface is used to create an automated adaptive ML system, which is initialized using pre-processed public datasets and *a priori* domain knowledge. A fundamental and progressive mapping between humans/people and states of a machine is defined by the system, allowing one side to automatically label the other.

1.2.8 User Interface and Experience (UI/UX)

Designing user interfaces with respect to the current technological advancements presents additional challenges, such as the need to be highly user-friendly, accommodate an increasing number of manufacturers, and effectively support a wider range of users. The advent of cyber-physical systems makes these difficulties worse. The incorporation of "smart" components results in systems that are more adaptive and versatile, but it also increases technological complications. One of the major challenges in user interface creation is to consequently decrease the apparent system complications with respect to the user.

1.2.9 System Feedback Loops

System feedback loop refers to the ongoing cycle of perception, judgment, and action involving interactions between the physical (environment or device), human, and cyber (computational) components. In combination with physical and computational systems, networks and embedded processors keep an eye on and manage these processes *via* feedback loops that allow for reciprocal communication. In CPHS, system feedback loops are closed-loop control mechanisms that measure the system's outputs and utilize them as inputs to affect subsequent behavior. Humans are involved in addition to machines and sensors. The fundamental forms of system feedback loops in CPHS are human-to-system feedback, system-to-human feedback, system-to-system feedback, and human-in-the-loop control.

1.2.10 Trust and Ethics

Collection and monitoring of data is one of the main ethical issues with CPHS. Sensors and other data-gathering tools are frequently used by CPHS to keep an eye on and manage the physical environment. This gives rise to concerns regarding user privacy and the possibility of data breaches. Because of the possible effects on people, organizations, and society at large, the creation and application of CPHS present serious ethical questions. In order to make sure that CPHS are created with the welfare, security, and privacy of people as their top priorities, ethics are essential. The following elements demonstrate the significance of ethics in CPHS development:

- Maintaining CPHS's dependability and safety
- Preserving user security and privacy
- Avoiding prejudice and discrimination
- Encouraging openness and responsibility
- Reducing possible adverse effects on society

1.3 Key Features of CPHS

CPHS differ from the traditional CPS through including humans as an essential part, whose important features are listed below (Figure 1.3).

1.3.1 Reactive Computation

Reactive computation in CPHS responds to inputs and events as they happen in real-time, allowing it to continuously interact with its surroundings. It is a computational paradigm in which the system continuously reacts in real time to outside inputs or events. Reactive systems are continuous, event-driven, and time-sensitive; they "react" to changes in their surroundings, in contrast to standard computation, which accepts input, processes it, and then stops.

1.3.2 Network Connectivity

The capacity of a CPHS device to get connected to the Internet is known as network connectivity. This can be accomplished by a translated connection,

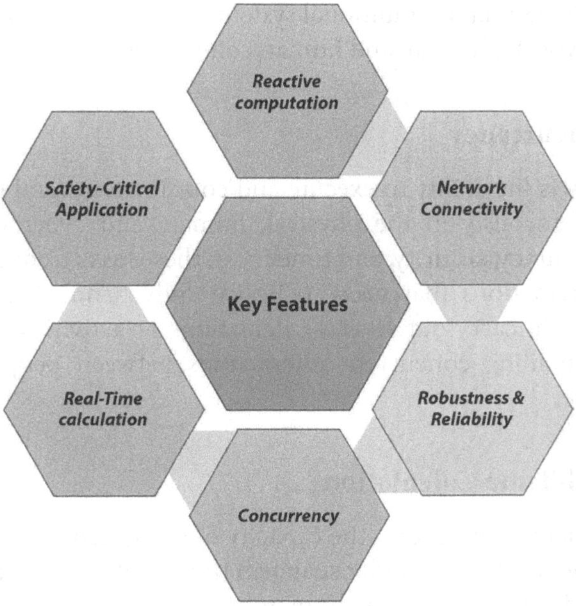

Figure 1.3 Key Features of CPHS.

which uses a private addressing scheme with a server that converts between public and private IP addresses. Every device is considered to have its own IP address when using a routed connection or shared Internet access.

1.3.3 Robustness and Reliability

The ability of the system to continue operating and performing under unpredictable, variable, or unfavorable circumstances, such as changes in the environment, sensor noise, human variability, or partial system failures, is referred to as robustness. In CPHS, robustness denotes the ability of the system to withstand disruptions, ambiguities, or mistakes in its physical, human, and cyber components without suffering appreciable performance or safety degradation.

On the other hand, reliability denotes the degree to which a system reliably carries out its projected purpose across cyber, physical, and human components over a predetermined period of time and under predetermined conditions, despite errors, delays, or human variability. It denotes the capacity of the system to operate accurately, consistently, and safely for a prolonged time period, even when faced with unforeseen circumstances, human variables, or uncertainties. Ensuring reliability becomes more difficult in CPHS than in traditional systems due to the close integration of computer (cyber), physical, and human components.

1.3.4 Concurrency

Concurrency is the ability to execute and coordinate several tasks or processes simultaneously in the physical, human, and cyber domains. To guarantee accuracy, security, and timeliness, these tasks frequently include real-time interactions that need to be carefully handled. It is the system's capacity to carry out several calculations or activities concurrently, frequently entailing concurrent interactions between people, software, and hardware.

1.3.5 Real-Time Calculation

Real-time calculation denotes the capacity of the system to process inputs and then generate outputs under stringent time limits, allowing for prompt decisions and actions involving human, physical, and cyber components. In CPHS, real-time calculation refers to the execution of calculations or control actions that need to be finished within a specified time frame to guarantee proper and secure operation of the system. It is based on actual facts from the real world; CPHS systems may make dynamic decisions thanks to their real-time calculation capabilities.

1.3.6 Safety-Critical Application

Systems that have the potential to fail and cause fatalities, serious injuries, or significant environmental or property damage are known as safety-critical applications. These systems in CPHS require very high dependability, robustness, and real-time responsiveness and require close coordination between cyber (software), physical (devices), and human components. It requires a safe operation in which any delay or malfunction, whether caused by the technology or a person, could have dire repercussions. With regard to CPHS applications, system safety takes precedence over system development and performance.

1.4 Application Areas of CPHS

A wide space for innovation and development across a variety of businesses is available through the integration of CPHS. It facilitates communication between digital and physical elements, which advances sectors like manufacturing, transportation, healthcare, and energy. CPHS may come across in order to better comprehend the possibilities and capabilities in a variety of fields, including robotics, medical devices, smart grids, building management systems, Industrial Internet of Things (IIoT), operational technology, smart transportation systems, and so on (Figure 1.4). Since CPHS has recently increased across all sectors and industries, it's critical to comprehend how to safeguard them due to a larger attack surface or exposure opportunities. By improving the networks' functionality, autonomy, dependability, safety, and usability, the increasing utility of smart cyber-physical human systems seeks to accelerate the deployment of large-scale systems.

CPHS has widespread use cases in the manufacturing sector, enabling more intelligent operations and revolutionizing conventional procedures. Predictive maintenance is one of the main uses. IoT devices and sensors keep an eye on equipment conditions all the time, anticipating failures before they happen. This reduces maintenance costs and downtime because repairs are only planned when necessary. Real-time observation

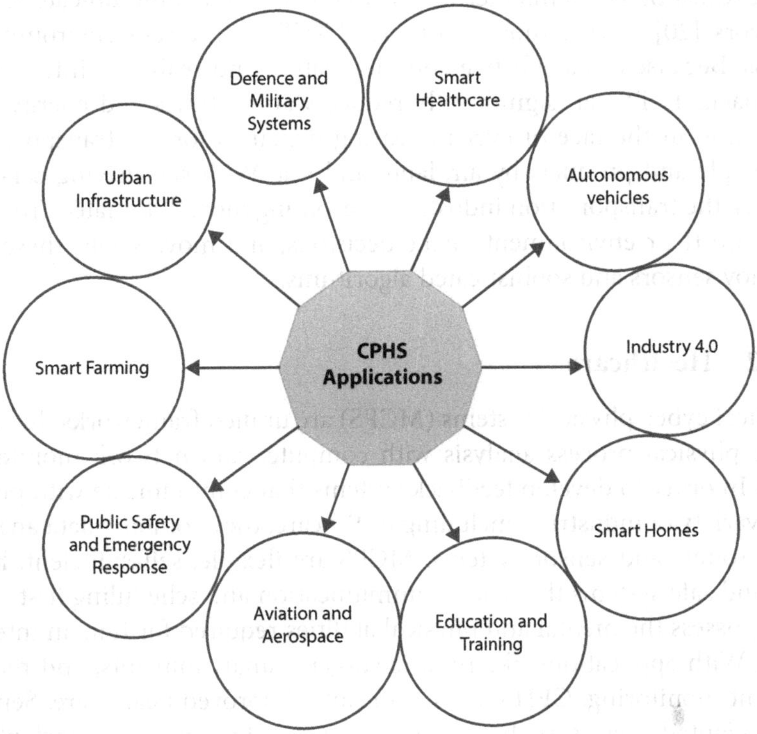

Figure 1.4 Application areas of CPHS.

and control are another essential use case [16]. Production lines can be continuously monitored with CPS, which offers real-time data to enhance workflows and raise product quality. This guarantees that any departures from the standard are promptly found and fixed, upholding high standards and cutting down on waste. Energy management, healthcare, education, agriculture, smart buildings, smart manufacturing, and smart transportation are the transdisciplinary areas of CPHS where it can be implemented [17–19]. The fields of monitoring civil infrastructure, robotics, decision-making, process control, diagnostics, and military are among the other intriguing CPHS application areas.

1.4.1 Transport Systems

Transportation cyber-physical systems (TCPS) is a term developed from CPS that specifically refers to the use of CPHS in the context of transportation systems. There are three categories of TCPS, namely, (1) infrastructure-based, (2) vehicle-infrastructure coordinated, or (3) vehicle-based. To improve the efficiency and fidelity of the transportation process, these three kinds of TCPS interact with actual items and communicate *via* the network [20]. Furthermore, widespread TCPS use lessens environmental stress. Because of its effective communication and real-time information feedback, TCPS can significantly reduce waiting times and energy consumption in the face of ever-increasing demands for the transportation of people and products by air, land, and sea. With self-driving cars and trucks, the transportation industry is becoming more automated. To comprehend their environment, make decisions, and move safely, these cars employ sensors and sophisticated algorithms.

1.4.2 Healthcare

Medical cyber-physical systems (MCPS) are unified frameworks that integrate physical process analysis with computer and network monitoring [21]. In order to develop feedback systems that communicate with people in a variety of industries, including healthcare, they combine data analysis with robots and sensor systems. MCPS are flexible, self-sufficient, helpful, and safe systems that place communication and scheduling first. They also possess the mental and physical abilities required for human interaction. With applications like robotic surgery, smart implants, and remote patient monitoring, CPHS has significantly improved healthcare. Sensors in implanted devices, such as insulin pumps and pacemakers, track physiological characteristics and provide tailored treatment.

1.4.3 Smart Cities, Homes, and Farming

Globally, a sizable portion of the population is moving towards metropolitan areas. To improve the well-being of their citizens, nations are pursuing smart city initiatives. The foundation of smart city configurations is CPHS. Nearly every system component in the ecosystem of a smart city uses them. On the other hand, a smart home is made up of many smart devices that automate the house and provide precedences, like comfort, efficient energy, safety, and remote control. CPHS are accelerating the development of smart homes by bridging the gap between the physical and digital worlds. One of the key elements of smart homes that encourages users to reduce their energy usage is energy efficiency.

The average household uses a lot of energy when its appliances are not in use and are still drawing power. CPHS is essential for power management, which lowers power consumption in smart home implementations. Similarly, the agricultural sector has become a crucial area of study in recent decades due to factors like resource constraints, rising demand, and climate change. To build on crop pliability and production efficiency, farmers have been forced to embrace innovations and new technologies. Technological breakthroughs like artificial intelligence (AI), digital twins (DT), and machine learning (ML) are essential to develop CPHS applications related to smart agriculture and demonstrate their benefits.

1.4.4 Emergency System

CPHS are used to maximize accurate results of disasters from a human standpoint, which includes emergency management systems (EMS), a significant example. CPHS are also used to maximize Quality of Service (QoS) besides primary assisting people with respect to safety, health, and survival as well as preservation of property, nature, and important organizations. EMS makes use of sensor networks, networked decision nodes, and relationships with evacuees and first responders [22]. Results will depend on how well the ICT system functions under pressure. Using expertise for EMS has a few drawbacks, such as malfunctions and malevolent attacks.

1.4.5 Education

CPHS revolutionizes the area of education through fusing computation, human interaction, and physical systems by producing intelligent, responsive, and adaptable learning environments. Smart classrooms, personalized learning, enhanced accessibility, remote and blended learning, data-driven decision-making, collaborative learning, and safety and security in educational environments are some of the fundamental functions and uses of CPHS. By creating more intelligent, flexible, inclusive, and effective learning environments, CPHS are transforming education. Their combination promotes fair access to high-quality education, individualized learning pathways, and holistic development.

1.5 Overview of the Book

This book describes CPHS's learning-based and AI-driven pricing algorithms, which are a revolutionary step in the development of dynamic, effective, and user-centric ecosystems. In settings where pricing techniques have a direct impact on system efficiency, user participation, and resource allocation, it also encompasses CPHS integration, physical processes, human interactions, and cyber technologies. This book incorporates the role of pricing in CPHS, including how it can optimize resource allocation, encourage human behavior, improve system efficiency, and facilitate market-driven operations. It covers fundamentals, nature, characteristics, and training models for CPHS, including real-time case studies of application-oriented CPHS whose flow and overview are illustrated in Figure 1.5. Additionally, it examines the developments in CPHS, which incorporates DNN, IoT, digital twin, IIoT, and human behavior modeling for use in

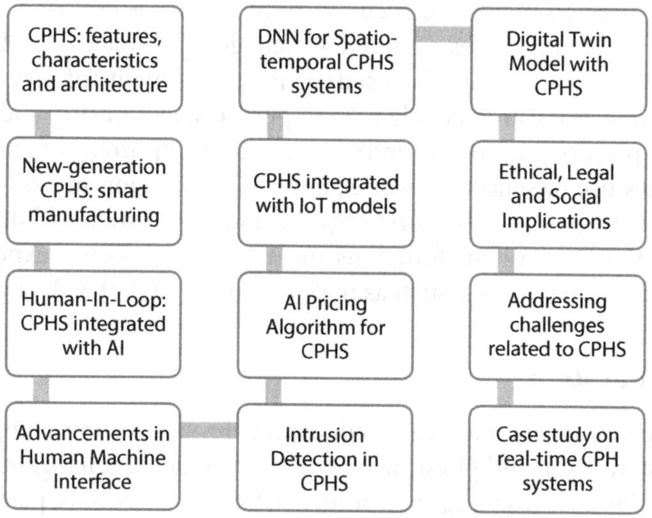

Figure 1.5 A brief overview of the chapters of the book.

fields such as autonomous vehicles, robotics, healthcare, 6G communication, and more.

The primary goal of this book is to cover new technological developments and applications in the framework of intelligent systems of the future that use CPHS. In order to create CPHS systems for any target application domain, it clarifies some technological features of Supervisory Control and Data Acquisition (SCADA), learning, AI pricing algorithms, and advanced HMI interfaces. Complex and uncertain situations can be solved by CPHS, which can also learn, gain cognitive skills, use knowledge more effectively, and create a hybrid method known as human-machine intelligence. It also examines the value and relevance of CPHS, which makes it possible to gather, analyze, and regulate data in real-time, resulting in automated and more efficient manufacturing processes. This book delves deeply into the cutting-edge HMI technologies that are essential to the effective deployment of cyber-physical systems, improving responsiveness, efficiency, and usability.

Systems known as SCADA are essential for keeping an eye on and managing vital infrastructure and industrial operations. SCADA networks must handle particular security, real-time responsiveness, and human safety issues when used in CPHS, where physical elements, cyber systems, and human contact all come together. For CPHS-related applications, this book covers the technical elements, difficulties, and solutions for deploying efficient intrusion detection systems (IDSs) within SCADA networks. Additionally, this book discusses the difficulties and problems associated with cyber-physical security systems and data privacy for a variety of real-time applications. In a similar vein, this book will discuss the advantages and disadvantages of combining digital technology, human behavior, and physical systems as a single, integrated platform to direct researchers, academicians, and developers.

1.6 Summary

Cyber-physical human systems are sophisticated systems that combine human interaction, computational operations, and physical elements to create environments that are intelligent, flexible, and context-aware. Real-time sensor data collecting, responsive actuation, computing, communication, and human-in-the-loop control methods that facilitate decision-making and system flexibility are essential components of CPHS. This chapter covered the fundamentals, basic architectures, perspectives, and application areas of CPHS, including an overview of the upcoming chapters in this book. CPHS has the possibility to entirely transform how people use technology in the future, resulting in more intelligent, user-friendly, and human-centered systems in every industry. CPHS will become increasingly autonomous, adaptive, and context-aware as AI, ML, edge computing, and sophisticated sensor technologies develop further. This will allow systems to react to human behavior, emotions, and requirements in real time with intelligence. Additionally, CPHS will broaden its scope to include collaborative human-robot interaction, in which humans and robots coexist in sectors such as manufacturing, agriculture, and space exploration. In order to guarantee that CPHS develops in ways that are safe, inclusive, and consistent with human values, ethical issues, privacy, and system trust will become key study topics.

AI disclosure statement: The author confirms that no artificial intelligence (AI)-generated content or AI-assisted writing tools were used in the preparation of this chapter (including summaries and the abstract). All text, analysis, and conclusions are the result of the author's original work.

References

1. Samad, T. and Annaswamy, A.M. (Ed.), *The Impact of Control Technology*, IEEE Control Systems Society, United States, 2011.
2. Khargonekar, P.P., Cognitive cyber-physical systems: Cognitive neuroscience, machine learning, and control. *In Proc. American Control Conf.*, United States, 4757, 2020.
3. Liu, Z. and Wang, J., Human-cyber-physical systems: concepts, challenges, and research opportunities. *Front. Inf. Technol. Electron. Eng.*, 21, 1535, 2020.
4. Khater, H.M., Empowering Healthcare with Cyber-Physical System a Systematic Literature Review. *IEEE Access*, 12, 83952, 2024.
5. Hu, L., Xie, N., Kuang, K., Zhao, K., Review of Cyber-Physical System Architecture. *In Proc. 15th Int. Sym. on Object/Component/Service-Oriented Real-Time Dis. Comp. Workshops*, China, 25, 2012.
6. Wang, Y., Vuran, M.C., Goddard, S., Cyber-physical systems in industrial process control. *ACM SIGBED Rev.*, 5, 1, 2008.
7. Acquaah, Y.T. and Kaushik, R., Normal-only Anomaly detection in environmental sensors in CPS: A comprehensive review. *IEEE Access*, 52, 1, 2008.
8. Sharma, A., Rathee, G., Kumar, R., Saini, H., Varadarajan, V., Nam, Y., Chilamkurti, N., A Secure, Energy and SLA-Efficient (SESE) E-Healthcare Framework for Quickest Data Transmission Using Cyber-Physical System. *Sensors*, 19, 2119, 2019.
9. Narayanan, S.K., Dhanasekaran, S., Vasudevan, V., A shared computational model using distributed processing in a CPS enabled environment. *In Proc. Third Int. Con. on Inventive Research in Comput. App.*, India, 3, 841, 2021.

10. Yuan, Y., Tang, X., Zhou, W., Data driven discovery of cyber-physical systems. *Nat. Commun.*, 10, 4894, 2019.
11. Mosqueira-Rey, E., Hernández-Pereira, E., Alonso-Ríos, D., Human-in-the-loop machine learning: a state of the art. *Artif. Intell. Rev.*, 56, 3005, 2023.
12. Ray, A., Autonomous perception and decision-making in cyber-physical systems. *In Eigth Int. Con. on Com Sci and Edu.*, Sri Lanka, 8, 10, 2013.
13. Jawhar, I., Al-Jaroodi, J., Noura, H., Mohamed, N., Networking and Communication. *In Cyber-physical Systems. In 37th Int. Con. on Distributed Comput. Sys. Workshops*, 37, 75, 2017.
14. Fink, G.A. and Edgar, T.W., *et al.*, Overview of Security and Privacy in Cyber-Physical Systems, *in Security and Privacy in Cyber-Physical Systems*, Song, H., Fink, G.A., Jescheke, S. (Ed.), pp. 1–23, Wiley, United States, 2017.
15. Madni, A.M., Sievers, M., Madni, C.C., Adaptive Cyber-Physical-Human Systems: Exploiting Cognitive Modeling and Machine Learning in the Control Loop. *INCOSE Int. Symp.*, 28, 1067, 2018.
16. Velu, A. and Thangavelu, M., Ontology based Ocean Knowledge Representation for Semantic Information Retrieval. *Comput. Mater. Contin.*, 70, 4707, 2022.
17. Chen, H., Applications of Cyber-Physical System: A Literature Review. *J. Ind. Inf. Integr.*, 2, 1750012, 2017.
18. Achar, S., Faruqui, N., Whaiduzzaman, M., Awajan, A., Alazab, M., Cyber-Physical System Security Based on Human Activity Recognition through IoT Cloud Computing. *Electronics*, 12, 1892, 2023.
19. Velu, A. and Thangavelu, M., Hetero-GCD2RDF: An Interoperable Solution for Geospatial Climatic Data by Deploying Semantic Web Technologies. *Wirel. Pers. Commun.*, 117, 3527, 2021.
20. He, Y., Jolfaei, A., Zheng, X., Cyber-Physical Systems in Transportation, *in: Intelligent Cyber-Physical Systems for Autonomous Transportation*, Garg, S., Aujla, G.S., Kaur, K., Hassan Ahmed Shah, S. (Ed.), pp. 76–98, Springer, Cham, 2022.
21. Shaikh, T.A., Rasool, T., Verma, P., Machine intelligence and medical cyber-physical system architectures for smart healthcare: Taxonomy, challenges, opportunities, and possible solutions. *Artif. Intell. Med.*, 146, 102692, 2023.
22. Gelenbe, E., Gorbil, G., Wu, F.J., Emergency Cyber-Physical-Human Systems. *In Twenty - First Int. Conf. on Comp. Comm. and Net.*, Germany, 21, 7, 2012.

2

Features, Characteristics, Architecture, and Future Potential of Cyber-Physical Human Systems

B. Srilatha[1*], Raghu Ramamoorthy[2], Mary Francy Joseph[3] and Anitha Velu[4]

[1]*Department of Electronics and Communication Engineering, T. John Institute of Technology, Bangalore, Karnataka, India*

[2]*Department of Computer Science and Engineering, The Oxford College of Engineering, Bangalore, Karnataka, India*

[3]*Department of Electronics and Communication Engineering, The Oxford College of Engineering, Bangalore, Karnataka, India*

[4]*Department of Electronics and Communication Engineering, Sri Sairam College of Engineering, Bangalore, Karnataka, India*

Abstract

This chapter covers information handling methods. This combines a physical process with a mathematical problem resolution activity. This is done by sensors. It will monitor and regulate processes at the same time. This system interconnects physical and nonphysical worlds. It makes human beings and machinery work together. Cyber-physical human systems (CPHS) sense and analyze activities while incorporating data protection methods. It can do synchronous data processing for analyzing the physical world operations. It can do complex analysis of data using various methods. This system is reliable and robust. It makes use of artificial intelligence and machine learning to automate processes and adapt to changing conditions, thereby improving system efficiency. As with most systems, CPHS systems are organized in multi-layered architectures, integrating cyber, physical, and human components. It is anticipated that the advancement of CPHS will drive cross-industry innovation, improving the collaboration between humans and machines.

***Keywords*:** Cyber-physical layers, decision support system, human-centric system, cloud-edge model, personalized assistance

**Corresponding author*: bsrilatha49@gmail.com; ORCID: 0000-0002-7721-3544

Anitha Velu, Prasanth Aruchamy, Raghu Ramamoorthy, Rajesh Kumar Dhanaraj and Manjula Arunraj (eds.) Cyber Physical Human Systems: Emerging Trends and Applications, (21–36) © 2026 Scrivener Publishing LLC

2.1 Introduction

Cyber-physical human systems are a combination of technology, hardware, and people [1]. The main difference between cyber-physical systems (CPS) and CPHS is that CPHS also incorporates humans. Here, humans actively interact with both cyber and physical components [2]. These systems interact with data processing units and machinery. Also, it is connected to human elements [3]. So, it helps for safe and efficient operation for the needs of humans [4]. As a result, it gained popularity in diverse fields. For example, transportation, manufacturing, health care, robotics, and emergency response [5]. CPHS helps in autonomous systems that work well in unpredictable situations. This is because of the computational intelligence, by including the human elements. It enables the systems to work in technologically vulnerable areas that are connected to humans. CPHS works differently from CPS because it works with reasoning skills and thoughtfulness. This is very helpful in real-time challenges that will increase safety and reduce performance-related issues [6].

2.2 Properties of Cyber-Physical Human Systems

CPHS has a special feature to adapt to real-time situations. It will help to work with advanced abilities that are sensitive and adaptive to different environments [7]. In these humans and machines are intelligently connected so that there will be easy problem identification. This skill enables them to respond to dynamically changing inputs. The input from automated information and the experiences of humans enhances integrated learning between humans and machines [8]. There are 4 main layers: the physical layer that consists of actuators, sensors, and devices; the next layer is the cyber layer, which comprises data; the third is AI and cloud units; and the fourth layer is the human layer. In the human layer, there are decision-making units. The interface layer connects humans and machines [9]. This layering in CPHS is suitable for manufacturing, the health field, building infrastructure, and disaster management. One important consideration is the integration of machines' reasoning with humans [10]. This means that CPHS is going to make wonders in future engineering and technology.

CPHS is essentially self-adaptive and user-friendly, which understands human needs. It provides safety, durability, and, moreover, an intelligent premise for industries. CPHS helps to integrate various departments of engineering and human systems. It interconnects digital systems and physical systems with human features. This will make coordination between them [11]. This makes CPHS reliable and robust. It can work in all unpredictable situations. They can be made to work from small-scale to large-scale human-oriented systems [12]. CPHS are based on social and ethical factors. Safety, accountability, and accessibility are important considerations [13].

The architecture of CPHS is generally organized into a set of interconnected layers that ensure fluid interaction among the system's key elements. A modern intelligent system is structured across multiple interconnected layers, each playing a crucial role. The physical layer includes hardware components such as sensors, actuators, and machines that interact directly with the physical environment. The cyber layer manages and processes information using computational platforms, artificial intelligence, and communication technologies. The human layer involves users, operators, or decision-makers, supported by models of human cognition and physiological behavior to ensure effective interaction. The interaction layer facilitates smooth and intuitive communication between humans and machines through control interfaces, augmented or virtual reality, and user-centered design. Finally, the decision and control layer integrates data from all other layers to enable real-time decision-making and efficient execution of system responses [14].

Looking ahead, CPHS are poised to revolutionize various sectors by enabling closer and more intelligent cooperation between humans and machines. In smart manufacturing, collaborative robots and humans will work side by side, adapting in real time to process changes and human feedback. Autonomous systems, including self-driving vehicles and unmanned aerial systems, will become more human-aware, responding to emotional cues and intent. In healthcare, the synergy between artificial intelligence (AI)–based diagnostics and human expertise will allow for more tailored, patient-specific treatments. Disaster response operations will benefit from improved coordination between human teams and autonomous systems, increasing both safety and operational effectiveness in chaotic environments. Moreover, advancements in cognitive systems, such as brain-computer interfaces, will deepen the connection between humans and machines, opening new frontiers in human-computer interaction.

2.3 Features of Cyber-Physical Human Systems

CPHS marks a significant evolution in intelligent system design by seamlessly combining digital computing technologies, physical hardware, and human interaction within a unified, interactive framework. In contrast to traditional CPS, which primarily focus on the integration of computational and physical elements, CPS place humans at the core of the system's operation. In this case, human elements are actively involved in synchronous decision-making. It also helps for a system to adapt through the response and collaboration of machines and humans. This enhanced technical ability of humans imparts stability and self-governance. CPHS have many applications in the areas of manufacturing, health, and automated transport. There should be smooth interaction between all the layers [15]. The data processing is done by different algorithms of software, artificial intelligence, and, finally, by digital testing. The physical components are sensors, actuators, machines, and robotics, which interact with the human layer.

CPHS analyzes the needs and perceptions of human beings, and then it will systematically respond. Human factors play a major role in efficient CPHS. Collaboration between human, cyber, and physical layers is very much needed for the working of CPHS. The analyzed data is fed to data analytics algorithms for processing, and based on the results, actions will be taken. Here, the physical components are sensors and actuators, which are directly in contact with other layers. CPHS always works on human needs and priorities; hence, the system effectively addresses all human features. CPHS is efficient and versatile and gives a big role to human features. For example, artificial intelligence helps human beings to smartly operate machines for increased productivity and a safer environment.

2.3.1 Real-Time Data Collection and Adaptive Feedback

CPHS gather data from the environment and human interactions without interruption. A number of instruments are employed, and many technologies are utilized by current advanced intelligent systems to gather and interpret information from both their environment and users. Hence, active monitoring of physical conditions is possible for embedded sensors and Internet-of-Things (IoT) devices, which enables real-time awareness and responsiveness. Through the capture of physiological and behavioral data, cameras and biometric systems can identify user interactions and states. Besides interactive interfaces, record maintenance, and users' actions, interests, and comments, the systems can change responses in real time while improving user experience. Real-time analysis of the data is carried out through AI algorithms and control systems, which then trigger appropriate responses or modify operational strategies as needed. This capability of dynamic processing enables CPHS to respond immediately to changing conditions or human inputs, making them extremely adaptable, flexible, and user-aware.

2.3.2 Autonomous Decision Support

Self-controlled or partial self-control is a crucial feature of cyber-physical human systems. Various factors play a role in the success of this process, and advanced tools in intelligent systems enable faster decision-making and flexibility. Artificial intelligence models can model the behavior of a system by examining data patterns. However, there are situations where optimization methods are necessary to determine which actions are optimal and which are not. Implementation of human-in-the-loop (HITL) mechanisms and automated processes in difficult situations allows for system decision-making by providing human review, adjustment, and override/oversight while also taking into account contextual judgment. This finds applications in complex systems where safety is a major concern, such as self-driving vehicles, robotics, and clinical support, but human involvement is needed to ensure safety and to avoid risk.

2.3.3 Human-Centric System Design

Interaction with humans and machines is an important aspect for CPHS, which tries to improve human-centric systems. Augmented reality (AR), natural language processing (NLP), and voice-controlled systems are used to achieve such centric systems. CPHS can find out the errors made by human beings. Human analytical ability distinguishes CPHS from other systems, where human reasoning can avoid failures that happen mechanically or electronically. Humans can determine the reason for the breakdown and resolve it; human beings can alter the way systems work for improved safety and robustness.

2.3.4 Collaborative Multi-Agent Ecosystem

CPHS integrates the efforts of human beings and equipment, which helps to accomplish objectives effectively. It is done by automated systems that do complex operations with the help of a control system to make human decisions. It is done by the joint endeavor of human beings, sensors, actuators, and different equipment. By simulating the behavior, we can convert the interaction of human beings to complex analytical components. For this, deep human reasoning is required, which leads to better coordination between human beings and machinery.

2.3.5 Ethical and Context-Aware Behavior

These automated systems are used to ease the efforts of human beings with the help of in-depth human reasoning and the ability to work in a multidisciplinary area. This reasoning technique differentiates CHPS from others in universal human values, ethics, and safety. Different cultures, social factors, ethical behavior, and emotional intelligence have an impact on the functioning of CPHS systems. These are based on laws like Asimov's robotic codes. It will prevent harmful effects and uphold moral values. CPHS can operate in sensitive environments such as healthcare, homes, and public spaces where human welfare and rights are highly valued due to these considerations.

2.4 Architecture of Cyber-Physical Human Systems

In applications such as autonomous transportation, Industry 4.0 smart manufacturing, robotic healthcare, and disaster management from smart cities, CPHS are an advanced combination of computational systems and physical components (like machines, devices, and environments) with human participants working in tandem. The CPS system concept is expanded upon, but humans are now considered indispensable and adaptive contributors to decision-making processes instead of inactive observers. Figure 2.1 illustrates the architecture of cyber-physical human systems.

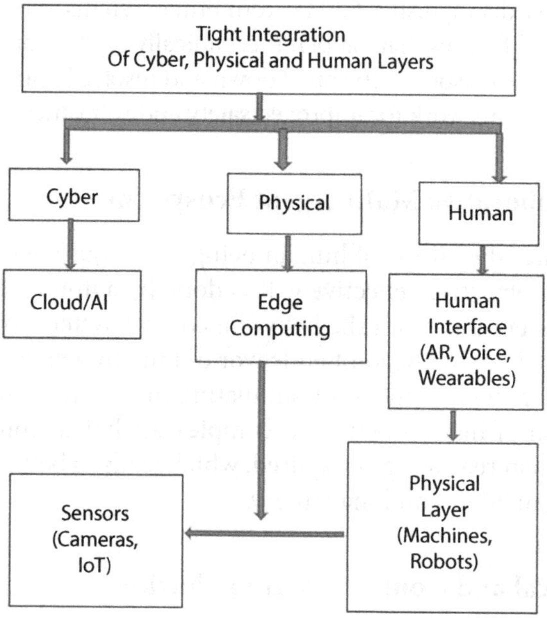

Figure 2.1 Architecture of cyber-physical human systems.

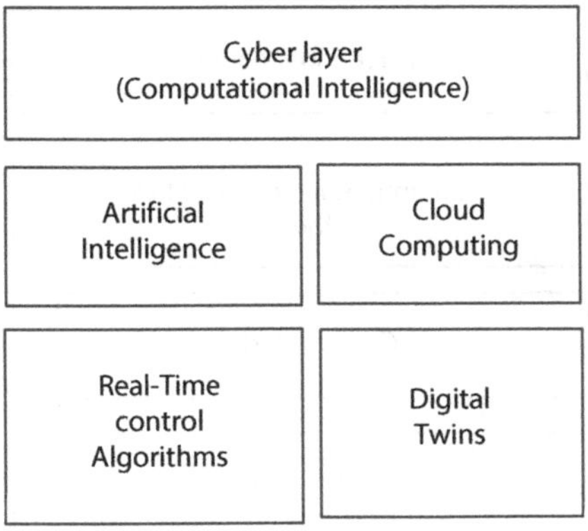

Figure 2.2 Cyber layer.

2.5 Cyber Layer (Computational Intelligence)

The cyber layer is the fundamental intelligence component of CPHS, facilitating intelligent decision-making, data handling, and communication between machines and humans. The "brain" of a system is frequently recognized as the brain, which comprehends data from physical sensors, incorporates user input, and directs actuators to execute tasks in real-world scenarios. The cyber layer is experiencing rapid growth as technologies like AI, machine learning (ML), edge computing, and cloud services continue to evolve. Figure 2.2 describes the cyber layer.

2.6 The Role of Computational Intelligence

In the CPHS layer, data from physical components like machines, sensors, and infrastructure, as well as human-centered inputs such as gesture/voice interactions and physiological signals, is primarily processed. Self-regulation and adaptability are the hallmarks of intelligent systems. Its aim is to deliver fast, high-quality responses by interpreting and processing the data from multiple sources in real time. This is particularly important since proactive optimization of the system can be achieved by anticipating predetermined actions, such as environmental changes or user behaviors. Furthermore, they possess the capacity to make autonomous choices or provide helpful counsel to human users, resulting in enhanced efficiency. The future functionality and performance of these systems can be continuously enhanced through continuous learning, as shown in Figure 2.3.

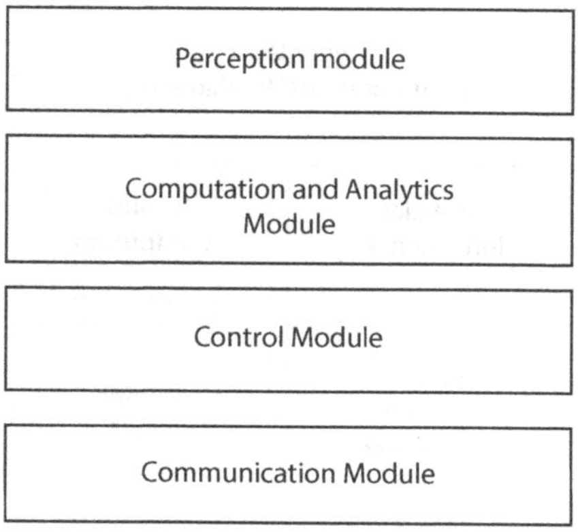

Figure 2.3 Components of the cyber layer.

2.6.1 Core Components of the Cyber Layer in CPHS

i) Data Collection and Processing

A sophisticated system relies on input from a range of sensors and interfaces to gather comprehensive real-time data. The raw data is processed and categorized to ensure precise and efficient analysis. Through real-time system tracking, the system is capable of finding irregularities at the beginning stage, thereby enhancing reliability and ensuring good performance.

ii) Perception and Contextual Awareness

Sensor data is transformed into actionable knowledge by intelligent systems through AI, resulting in smart and informed decision-making. They provide for advanced capabilities such as object recognition, activity tracking, and emergency detection across a wide range of domains.

iii) Intelligent Decision-Making and Strategic Planning

Decision processes in intelligent systems are guided by techniques such as rule-based logic, reinforcement learning, and probabilistic models. The situational context is analyzed by these systems to determine the most appropriate course of action in real time. Cooperative tasks require accurate, context-aware responses and mutual understanding to ensure effective collaboration between humans and robots.

iv) Adaptive Learning through Machine Learning

The ability to learn from previous data, user behavior, and dynamic environments enables intelligent systems to adapt over time and evolve environments. To improve accuracy and operational efficiency, this layer regularly updates its internal models. For example, a smart home can progress by identifying the interests of its inhabitants, optimizing lighting, temperature, and energy consumption to meet their needs.

v) Communication and System Coordination

Intelligent systems are designed to facilitate real-time communication between devices, users, and cloud infrastructures, ensuring smooth operation. Moreover, they manage synchronization, ensure fault tolerance, and implement robust data security measures to maintain the system's integrity and reliability across all parts.

vi) Incorporating Human Feedback

The cyber layer receives both direct input from humans (e.g., manual controls or voice commands) and indirect signals (e.g., eye movement, pulse rate, or facial expressions). This feedback helps fine-tune responses. For example, by analyzing employee fatigue using machine learning and biosensor data, the actions of a robotic assistant may be hindered.

2.7 Architectural Frameworks

2.7.1 Cloud-Based Architecture

Training complex AI models in centralized systems is advantageous because it requires minimal computational power and extensive data. The use of edge or fog computing results in faster and lower latency due to the implementation of data processing closer to its source, such as sensors or devices.

2.7.2 Hybrid Cloud-Edge Model

A balanced and efficient system is created by combining the advantages of cloud computing with edge computing. Highly demanding applications, such as autonomous driving and industrial automation, require high computational power and low-latency responses, making this architecture highly effective.

2.7.3 Key Challenges

Addressing the challenges of handling diverse data from humans and machines, ensuring transparency meets privacy requirements, and responding promptly to requests and changes is crucial. In systems, the adoption of human-centered design can be an effective way to convey moral values. Intelligent analysis, learning, and decision-making are made possible by the cyber layer.

2.8 Physical Layer

Figure 2.4 illustrates the physical layer, which is the basis of CPHS, and includes components like sensors, actuators, and infrastructure. It facilitates real-time engagement between the physical world, humans, and the cyber layer through data collection/conduction and command execution.

2.8.1 Key Components of the Physical Layer in CPHS

i) Sensors - Sensors play a crucial role in intelligent systems by gathering data from the environment and users to respond quickly. There are various types of sensors, such as environmental, motion, and physiological sensors; all require input from the system to perceive and interact with the external world in order to accurately assess human conditions.
ii) Actuators - These are the mechanical devices that perform operations in the physical world. This works on the basis of decisions made by the cyber layer and inputs from humans. It bridges the gap between machines and humans.
iii) Physical infrastructure - These are the hardware platforms for CPHS. For example, smart vehicles, smart gadgets, etc. The coordination between sensors and actuators is the main feature of CPHS.

2.8.2 Functions of the Physical Layer

The physical layer collects instantaneous data from sensors and actuators. It helps in fault-tolerant operation.

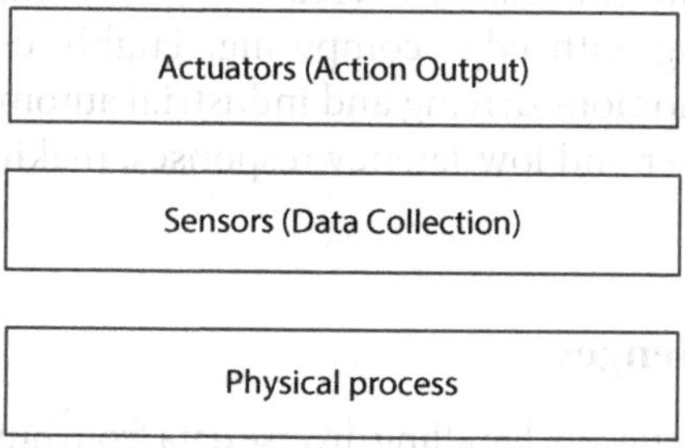

Figure 2.4 Physical layer.

i) Human-Physical Interaction

The use of smart devices is becoming popular now. These devices study the behavior of human beings by tracking the user's senses. This technology will provide safety, ease, and a customized experience.

2.9 Integration with the Cyber Layer

The cyber layer is integrated into the physical layer. Also, the physical connects the digital world and human beings. The cyber layer analyzes data that is collected from the cyber layer and predicts the behavior. The cyber layer collects data and then processes and takes action through actuators and other devices based on the outside world's context. The system remains dynamic, adaptive, and context-aware thanks to this two-way communication. The physical layer CPHS allows sensing and actuation in the physical world using sensors and actuators. It has to deal with issues such as precision, synchronization, hardware monitoring, and the safety of the end-users. As systems become more human-centered, this layer needs to be ever more flexible, reliable, and responsive.

2.10 The Human Layer: Cognitive and Decision-Making Role

The human layer separates CPHS from regular systems for adding human intelligence, emotions, ethical aspects, and reasoning abilities as depicted

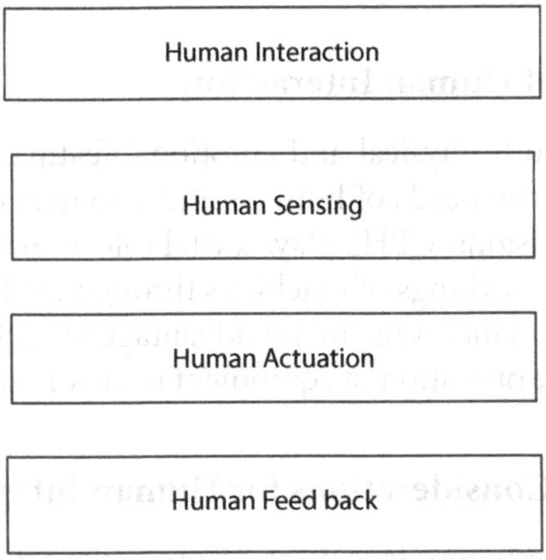

Figure 2.5 Human layer.

in Figure 2.5. Different from traditional cyber-physical systems in which automation and mechanical control are the primary focus, for CPHS, humans are dynamic thinking entities playing a role of interacting, influencing, and making decisions together with machines.

2.10.1 Core Functions of the Human Layer in CPHS

In CPHS, the human layer serves as a source of information and a decision-maker. It facilitates perception and awareness, serves moral judgment, and plays a necessity-taking (*b/c*) role as a necessary but non-sufficient condition for ethical decision-making. It also enables learning from machines through feedback transfer with humans and gives instant emotional/physiological input that influences the action of the system.

2.10.2 Cognitive and Decision-Making Elements

i) Cognitive Abilities

Cyber-physical human systems can be much more effective due to human cognitive capabilities. CPHS senses the data and interprets it through reasoning ability. For this emotional intelligence and ability, it is mandatory to make decisions in uncertain situations. This ability helps in problem-solving.

ii) Decision-Making based on Human-Machine Integration

Human-machine integration enables reasoning ability to work in risky situations. Human beings authenticate this data and overrule the mechanized decisions. This is applicable in self-driven vehicles, health care, and defense. Here, safety, moral rules, and decisions based on critical thinking are important.

2.10.3 Modes of Human Interaction

CPHS integrates both physical and emotional features of human beings. It helps to identify the needs of humans and also uses data analytics algorithms to make decisions. CPHS plays a vital role in matching the behavior of humans and the workings of machines through feedback loops. It leads to enhanced performance. One major advantage of CPHS is humans, who can intervene in the operation of equipment to ensure safety and reliability.

2.10.4 Design Considerations for Human Integration

Researchers and designers face a lot of problems while designing these systems, in which a major intricacy is human behavior, which behaves differently in each situation. Second is safety and reliability. Maintaining human values and ethics is also an important concern. The major advantage of CPHS is speed and data analytical ability, but maintaining integrity is a problem.

2.10.5 Collaborative Decision-Making

The human layer makes decisions based on inputs from human beings, like morality, emotional intelligence, and creativity. This is analyzed and translated using artificial intelligence, which helps in assessing the conditions of patients and also in the manufacturing area to produce smart products with automated responses.

2.10.6 Challenges in the Human Layer

There are several challenges in building human-machine collaboration. Different people have different ways of analyzing and reacting to the same situations, and also, thinking ability varies from person to person. This causes unpredictability because, for a specific situation, behavior patterns vary, but we cannot neglect or be confident in the results. A machine- and human-friendly environment should be created.

2.11 The Future Potential of Cyber-Physical Human Systems

2.11.1 Personalized Healthcare and Assisted Living

CPHS will be groundbreaking in the field of health because it can constantly assess patients; also, it can adaptively take actions. Smart health monitoring devices collect data on a continuous basis and send it to artificial intelligence-based data analytics systems. One of the main advantages is early disease detection and predicting forthcoming illness. The data analyzed will be sent to both patients and healthcare specialists for further action to take preventive measures. Consider an example of an old person who lives alone, and his/her vital parameters must be checked, and his/her relatives informed about his/her health conditions. Also, any malfunction in the device should be prevented. The most important are the parameters. According to the comfort of patients, it should be adjusted. CPHS can provide the most advanced solutions in the health sector to reduce hospital cases and increase the quality of life.

2.11.2 Industry 5.0: Human-Centric Manufacturing

CPHS is going to revolutionize human-machine interaction by integrating technology with it. This can provide a safe and automated environment. The major challenges include data and network security, transparency, and reliability. Also, modeling unpredictable behavior poses a big challenge. Handling large amounts of data from sensors is also a problem. It may raise ethical and legal issues, but in the future, its full potential may be utilized.

2.11.3 Autonomous and Connected Mobility

CPHS is ready to change transportation by enabling intelligent, human-conscious mobility. In addition to self-referential cars, future systems will explain human intentions and emotions. This will allow vehicles to make smart decisions based on context. Imagine cars that reduce speed for confused pedestrians or stop their driving style based on luggage stress. These systems will also support cooperation between drivers, passengers, and traffic authorities. This will create a safer and more flexible transport network.

2.11.4 Smart Environments and Urban Resilience

CPHS introduces human cognition, emotions, and decisions within system feedback loops to influence how we interact with technology and the environment. Sustainable by Design will create these ecosystems across key industry sectors and change how we design spaces by linking sensor networks, AI algorithms, and humans to decide what the space does. CPHS is revolutionizing homes and workplaces by creating intelligent spaces that learn from human behavior to optimize comfort, energy use, and safety. In urban planning, CPHS enhances disaster response by combining human responders, autonomous drones, AI systems, and IoT infrastructure.

2.11.5 Human-Machine Collaborative Decision-Making

Cyberphysical human systems' primary strength is their capacity to support intricate decision-making procedures. CPHS integrates machine learning, real-time analytics, and human ethical reasoning in domains like healthcare, finance, military operations, and crisis response. Rather than replacing humans, these systems act as cognitive partners. For example, AI may detect subtle patterns in medical scans while doctors apply contextual expertise, resulting in more precise and ethical outcomes.

2.11.6 Challenges and Research Directions

Cyber-physical human systems are transforming the relationship between humans and technology by creating safer and more intelligent environments. CPHS is still under research. The major challenges are network and data security, and ethical problems. The main research is to integrate humans and machines to create an automated environment. Obviously, the development of artificial intelligence will create a great innovation in this field.

2.12 Conclusion

CPHS synchronizes machines with human inputs and different technologies that lead to highly automated computational models that give importance to human beings. They combine the efforts of human beings and equipment for decision-making without delay. Efficient collaboration can be done by combining the functions of layers. The physical, cyber, and human layers ensure CPHS is reliable. The system helps to work with a strict time frame. These different types of protocols exist. It helps to maintain ethical values. CPHS facilitates building smart systems that are helpful in manufacturing, health care, etc. CPHS can also make a great transformation in the transportation industry. In the future, CPHS can make great changes in society.

AI disclosure statement: The author confirms that no artificial intelligence (AI)-generated content or AI-assisted writing tools were used in the preparation of this chapter (including summaries and the abstract). All text, analysis, and conclusions are the result of the author's original work.

References

1. Fraga-Lamas, P., Barros, D., Lopes, S.I., Fernández-Caramés, T.M., A cost-effective IoT thermal imaging safety system using mist and edge computing in Cyber-Physical Human-Centered Systems for Industry 5.0. *Sensors*, 22, 8500, 2022.
2. Sbaragli, A., Ghafoorpoor, P.Y., Thiede, S., A cyber-physical architecture to monitor human-centric reconfigurable manufacturing systems. *J. Intell. Manuf.*, 9, 211, 2025.
3. Rosen, D. and Choi, C., Research issues in the generative design of cyber-physical human systems. *J. Comput. Inf. Sci. Eng.*, 23, 7, 2023.

4. Mukhopadhyay, S., Liu, Q., Collier, E., Zhu, Y., Gudishala, R., Chokwitthaya, C., DiBiano, R., Nabijiang, A., Saeidi, S., Sidhanta, S., Ganguly, A., Context-Aware Design of Cyber-Physical Human Systems. arXiv preprint, 7, 132, 2020.
5. Werner, D., Martin, F., Alyssa, J.K., A References Architecture of Human Cyber-Physical Systems – Part III: Semantic Foundations. *ACM Trans. Cyber-Phys. Syst.*, 8, 116, 2024.
6. Smith, A., Jones, B., Lee, C., A human cyber-physical system for human-centered computing in seafaring. *J. Ambient Intell. Humaniz. Comput.*, 14, 7871, 2023.
7. Strobbe, C., Salas Cordero, S., Vingerhoeds, R., Valès, S., Towards a novel model-based multi-disciplinary design approach for human cyber-physical systems. *J. Manuf. Syst.*, 12, 1456, 2024.
8. Hoang, D.C., Sindhu, S.S., Krishnanand, K.R., Cyber-physical Infrastructure for Smart Building Applications towards Human Comfort and Energy Savings. *J. Dyn. Syst. Meas. Control*, 1, 12, 2021.
9. Damm, W., Hess, D., Schweda, M., A References Architecture of Human Cyber-Physical Systems – Part I: Fundamental Concepts. *ACM Trans. Cyber-Phys. Syst.*, 8, 32, 2024.
10. Sbaragli, A., Ghafoorpoor, P.Y., Thiede, S., A cyber-physical architecture to monitor human-centric reconfigurable manufacturing systems. *J. Intell. Manuf.*, 22, 1342, 2025.
11. Liu, Z. and Wang, J., Human-cyber-physical systems: concepts, challenges, and research opportunities. *Front. Inf. Technol. Electron. Eng.*, 21, 1535, 2020.
12. Fu, Z., Chao, C., Wang, H., Wang, Y., Toward the participatory human-centred community: an exploration of cyber-physical public design for urban experience. *IET Cyber-Phys. Syst.:. Theory Appl.*, 4, 209, 2019.
13. Li, T.N.I. and Bahsoon, R., A conceptual References model for human as a service provider in cyber-physical systems, based on SoA ontology model, arXiv preprint, 21, 164, 2021.
14. Yilma, B.A., Panetto, H., Naudet, Y., Systemic formalisation of Cyber-Physical-Social System (CPSS): a systematic literature review, arXiv preprint, 15, 128, 2021.
15. Zhou, J., Zhou, Y., Wang, B., Zang, J., Human-Cyber-Physical Systems in the Context of New-Generation Intelligent Manufacturing. *J. Eng.*, 5, 624, 2019.

3

Technological Aspects of Advanced Human–Machine Interface for Cyberphysical System Implementation

B. Srilatha[1*], Anitha Velu[2], Thanka Saranya C.[3] and Meghana Karri[4]

[1]*Department of Electronics and Communication Engineering, T. John Institute of Technology, Bangalore, Karnataka, India*
[2]*Department of Electronics and Communication Engineering, Sri Sairam College of Engineering, Bangalore, Karnataka, India*
[3]*Department of Electronics and Communication Engineering, The Vemana Institute of Technology, Bangalore, Karnataka, India*
[4]*Department of Computer Science and Engineering, University of Wisconsin, Madison, Wisconsin, USA*

Abstract

The advancement of human–machine interfaces (HMIs) plays an essential role in the successful implementation of cyberphysical systems (CPSs), as they greatly enhance interaction between humans and intelligent systems. The technological aspects of advanced HMI for CPS are covered in this chapter with a focus on multimodal interaction, sensor integration, augmented reality, virtual reality, and actual component feedback mechanisms. Artificial intelligence, machine learning, and edge computing combine to create adaptive and context-aware interfaces that enhance user experience and system efficiency. Secure communication protocols and privacy-preserving techniques help safeguard sensitive data, whereas real-time feedback mechanisms enable more responsive and dynamic control. The use of digital twins and haptic feedback technology offers immersive real-time monitoring and control capabilities, especially in critical applications. For efficient, intelligent, and safe human–CPS collaboration in modern industrial and societal contexts that lead to cyberphysical human systems, HMIs are essential. They can

**Corresponding author*: bsrilatha49@gmail.com; ORCID: 0000-0002-7721-3544

Anitha Velu, Prasanth Aruchamy, Raghu Ramamoorthy, Rajesh Kumar Dhanaraj and Manjula Arunraj (eds.) Cyber Physical Human Systems: Emerging Trends and Applications, (37–70) © 2026 Scrivener Publishing LLC

achieve seamless system operations, individualized interactions, and low-latency communication by combining these technologies.

***Keywords*:** Human–machine interfaces, Internet of Things, augmented reality, virtual reality, sensors and edge computing

3.1 Introduction

An essential part of cyberphysical systems (CPSs) is the human–machine interface (HMI), which makes it easier for humans and machines to communicate. HMIs have developed over time from simple control panels to complex artificial intelligence (AI)–powered systems that provide predictive analytics, adaptive user experiences, and real-time data processing. By leveraging cutting-edge technologies such as haptic feedback, augmented reality (AR), AI, and multimodal interfaces, advanced HMIs enhance usability, efficiency, and real-time decision-making in CPS applications [1, 2]. Furthermore, edge computing (EC) and the Internet of Things (IoT) have been combined to change the responsiveness of HMIs, making them an essential component of modern CPS applications. Technological developments across the industry have led to the development of CPSs. Such a system blends digital technology with the physical elements of a procedure. In addition to being a major force in technological progress, the CPS can act upon and expand shop floor components by applying computational principles. CPSs have been incorporated into the development of Industry 4.0 and the industrial IoT.

CPSs rely on HMI, which facilitates communication between humans and machines. From basic control panels to sophisticated AI-powered systems that support real-time data processing, adaptive user experiences, and predictive analytics, HMIs have evolved over time [3, 4]. Because they enhance the usability, efficacy, and decision-making potential of CPS applications, technological advancements such as haptic feedback, multimodal interaction, AR, and AI have fueled this growth. Active interfaces are no longer the standard for modern HMIs; instead, intelligent context-aware systems that adapt dynamically to user behavior and environmental factors have taken their place. HMI is now more broadly applicable in industries such as manufacturing, healthcare, transportation, and smart cities because of these developments [5].

In Figure 3.1, cyberphysical frameworks refer to the connection between handling frameworks and physical frameworks through advancements in communication sensors and actuators. Increased efficiency, safety, and intelligence in industrial and social applications are made possible by the

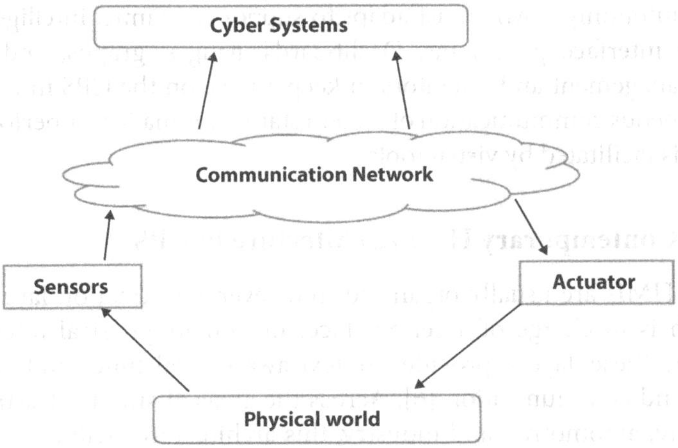

Figure 3.1 Cyberphysical system.

integration of HMI in CPSs, which is revolutionizing human interaction with automated processes. For any cyberphysical human system (CPHS) to respond dynamically to changes in the environment and human inputs, contemporary HMIs are increasingly incorporating predictive analytics, adaptive learning models, and real-time data processing. By bridging the gap between human cognition and machine intelligence, these interfaces guarantee seamless collaboration in applications such as autonomous vehicles, smart infrastructure, healthcare, and smart manufacturing. It also discusses emerging problems and possible directions for developing safer, more efficient, and easier-to-use HMIs for CPHS.

3.2 HMI in Cyberphysical Human Systems

HMIs are a vital component of CPHSs, serving as the interactive conduit between human users and the digital–physical infrastructure. HMIs need to evolve to support dynamic, intelligent, and intuitive interface paradigms as CPS facilitates more complex system behaviors, real-time decision-making, and increased autonomy. Using dashboards, gauges, graphs, and alarms, operators and management can monitor the CPS in real time. Visual tools make it easier to communicate system statuses, anomalies, or performance metrics instantly.

A key element of CPHSs is the HMI, which acts as an interactive link between users and the digital–physical infrastructure. As CPS enables increasingly complex system behaviors, real-time decision-making, and greater autonomy, HMIs must adapt to support dynamic, intelligent, and intuitive interface paradigms. Dashboards, gauges, graphs, and alarms allow management and operators to keep an eye on the CPS in real time. Instantaneous communication of system status, anomalies, or performance metrics is facilitated by visual tools.

3.2.1 Contemporary HMI Architecture in CPS

In CPS, HMIs are usually organized into several abstraction layers, each of which is in charge of a certain facet of human–physical interaction. Together, these layers provide context-aware, real-time, and intuitive control and communication [6]. Across the areas of smart infrastructure, healthcare, automotive, and industry, this architecture facilitates context-aware interactions, adaptive control, real-time monitoring, and predictive insights. Figure 3.2 shows the architecture of CPS. The fundamental layers of a contemporary HMI architecture in CPS are as follows:

(i) Perception Layer (Data Acquisition Layer)

The perception layer's primary job is to gather unprocessed data from sensors, cameras, microphones, and other input devices. Environmental sensors (such as temperature, pressure, and humidity), physiological sensors (such as eye tracking and electroencephalograms [EEGs] for brain–computer interfaces [BCIs]), and user input devices (such as wearables and touchscreens) are the elements that make up the perception layer. HMI's primary function is to allow the interface to perceive the user and the physical system, which serves as the foundation for interaction [7].

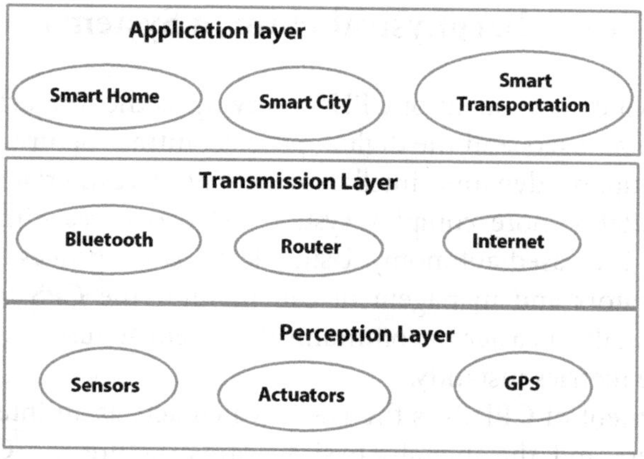

Figure 3.2 Layers of CPS architecture.

(ii) Information Exchange Layer (Presentation and Communication Layer)

The information exchange layer's primary job is to promote bidirectional information flow by transforming raw data into relevant visualizations, messages, and controls. Dashboards, warnings, status indicators, voice and gesture recognition modules, and aspects of virtual reality (VR) and AR make up the information exchange layer. HMI's primary function is to provide an interactive medium for human input and comprehension. By providing concise, contextual input, it facilitates situational awareness and decision-making.

(iii) Interpretation Layer (Semantic Layer/Cognitive Layer)

The interpretation layer's purpose determines intent and meaning by interpreting system context and user input. Natural language processing (NLP), user behavior modeling, and AI-based intent identification are the technologies utilized in the interpretation layer. The interpretation layer's function in HMI is that it closes the semantic divide between machine logic and human language and behavior, which allows for flexible and context-aware interaction.

(iv) Control and Command Layer

The purpose of the control and command layer is precisely to convert translated commands into control signals or system-level actions. Actuator controllers and command execution engines are parts of the control and command layer. Mechanisms for safety and override are also included. By changing parameters, starting tasks, or overriding automated functions, the control and command layer in HMI gives users direct control over the physical system.

(v) Feedback and Adaptation Layer

The feedback and adaptation layer's job is to respond quickly and adaptively to changes in the environment, user input, and system state. AI for dynamic user interface adaptation, reinforcement learning for user interaction patterns, user modeling, and personalization engines are the technologies utilized in the feedback and adaptation layers. Context-sensitive, configurable interfaces that adjust to the needs of the user and the system are the main features of the HMI feedback and adaptive layer, which enhances usability and user experience.

(vi) Security and Trust Layer

The security and trust layer's goal is to ensure that CPS and human users can communicate safely and dependably. Mechanisms used in the security and trust layers include secure logging, audit trails, biometric authentication, and role-based access control. The security and trust layer of HMI is primarily responsible for safeguarding user data, combating interface abuse, and promoting confidence in system autonomy.

(vii) Integration and Communication Layer

The integration and communication layer serves to facilitate seamless connectivity between the HMI and the underlying CPS components, edge devices, and cloud services. At the integration and communication layer are technologies such as data buses, middleware platforms, application programming interfaces (APIs), and communication protocols. In real-time CPS operations, the integration and communication layer of the HMI plays a crucial role in ensuring data consistency and interoperability across remote contexts.

3.3 Technological Aspects of Advanced HMI for CPS

Modern technologies are integrated into advanced HMIs for CPSs to facilitate smooth communication and cooperation between people and intricate systems. Across many industries, these technologies promote creativity, effectiveness, and improved user experiences. The few technical features are multimodal interaction in advanced HMI for CPS, sensor integration in advanced HMI for CPS, AR, and VR in advanced HMI for CPS [8, 9].

3.3.1 Multimodal Interaction in Advanced HMI for CPS

Multimodal interaction pertains to an HMI's capacity to support and manage various human input and output channels such as voice, gesture, touch, eye movement, and haptic feedback to improve the fluidity, reliability, and accessibility of interactions between humans and CPSs. It serves as a key aspect of cutting-edge HMI design, facilitating context-aware, user-adaptive, and resilient communication. To ensure effective interaction between individuals and machines, an HMI must be capable of processing and responding to an array of input and output modalities, including spoken language, touch, gestures, visual cues, and even physiological signals. This approach aims to mimic the natural way people communicate, which often involves combining several modes (for instance, talking while gesturing or gazing).

Multimodal interaction is not independent; it is closely linked with and enhances various other HMI technologies. In context-aware systems, it facilitates dynamic transitions or combinations of modalities depending on the user's state or environment. In the realm of AI, it leverages machine learning (ML) to analyze multimodal input patterns and improve responses. In EC, it guarantees the swift processing of multimodal signals with low latency. In cybersecurity, it enhances security through multimodal authentication methods, such as facial and voice biometrics. In visualization, it customizes visual feedback according to the input method, for instance, using eye gaze for attention-focused user interfaces. In augmented and VR systems, it merges gestures, voice commands, and haptic feedback to create an engaging interaction experience in virtual CPSs.

The core technologies behind multimodal interaction are sensor technologies, algorithms for sensor fusion, managers for multimodal dialogue, engines for context recognition, and feedback mechanisms. The sensors used are microphones, cameras, light detection and ranging (LiDAR), depth-sensing devices, inertial measurement units (IMUs), and EEG/electromyography. The algorithms for sensor fusion integrate various inputs for a unified interpretation. The managers for multimodal dialogue synchronize interaction modes in real time. The engines for context recognition deduce user intent and circumstances to prioritize inputs. Feedback mechanisms provide output through visual, auditory, and tactile channels.

(i) Common Modalities in HMI for CPHS

The visual modality involves input through facial expressions, eye tracking, and gesture recognition, whereas the output comprises visual alerts, AR, and graphical user interfaces (GUIs). In the auditory modality, input is gathered *via* speech recognition, voice commands, and environmental sound analysis, with the output consisting of speech synthesis, auditory notifications, and sound cues. The tactile modality utilizes touchscreens, force sensors, and wearable haptic gloves for input and provides output *via* vibration feedback, force feedback, and tactile displays. In the kinesthetic modality, body movement is monitored using IMUs and motion capture systems as input. The outputs consist of robot movement replication and feedback from exoskeletons.

In the olfactory and gustatory modalities (which are less common), chemical sensors (which are rare) act as input. The outputs include scent emitters, often used in training or simulation settings. Within the physiological modality, inputs include BCIs, heart rate monitors, skin conductance measures, and EEGs/EMGs. The system's behavior is adjusted based on the emotional or cognitive state of the user, serving as the output. Accessibility, context awareness, robustness, and naturalness are the benefits of multimodal interaction in CPS. Reliability is increased by robustness, which means redundancy across modalities. Being natural makes interaction more instinctive and human-like. Context awareness is the ability to better comprehend context by combining various sensory information. Accessibility allows people with varying abilities and preferences to be supported.

(ii) Technologies Enabling

The key technologies enabling multimodal interaction in human–machine interaction for CPSs are sensor technologies, fusion and processing technologies, interface and output technologies, communication technologies, AI, and cognitive systems. The ability to perceive various input modalities is made possible by sensor technologies. These make use of cameras for RGB (red–green–blue), depth (Kinect, RealSense, etc.), and facial, gesture, and posture detection. For sound localization and voice recognition, they make use of microphones and microphone arrays. For capacitive, resistive, and force sensors for touch and pressure input, they make use of touch sensors. They track gestures and body movements using motion sensors and IMUs. They record muscle and cerebral activity using BCI sensors (EEGs/EMGs). Eye-tracking sensors are used for attention modeling and gaze detection. Models that handle data from various modalities (such as transformers and multimodal encoders) take advantage of multimodal ML. They interpret written or spoken commands using NLP. They make use of computer vision for object detection, facial analysis, and gesture recognition. For noise reduction, speaker identification, and speech detection, audio signal processing is used.

Interface and output technologies provide users with output *via* a variety of channels. They make use of head-up displays (HUDs), AR/VR headsets, and GUIs [10]. They make use of force feedback devices (such as wearables and haptic gloves) and haptic actuators for vibration motors. Text-to-speech systems use voice synthesis engines to produce audio. They make use of AR/VR systems to create immersive settings that incorporate haptic, visual, and aural feedback. They make use of projection mapping to directly display visual data on tangible surfaces. Communication technologies allow the HMI to be integrated with the network and CPS. They take advantage of EC to enable multimodal processing in real time close to the source. They transmit sensor data with low latency using IoT protocols.

They support low-latency and high-bandwidth connectivity by using 5G and 6G. AI and cognitive systems promote perceptive, flexible communication. To manage multimodal discourse, they make use of voice assistants and conversational agents, such as Google Assistant and Amazon Alexa. They adapt interactions according to the user's emotional state by using emotion recognition systems. Context-aware systems are used to customize interactions by understanding user intent and environmental context.

(iii) Design Considerations

To guarantee efficient, user-friendly, and secure interaction, a number of design factors need to be considered while creating multimodal interaction for human–machine interaction in CPSs. Select the appropriate modality for the task. For instance, voice commands might work well when driving, but not in noisy settings. The user environment is important; when hands are occupied or there is a lot of glare, visual feedback might not be the best option. Use modalities, such as gestures and words, which enhance each other. Redundant modalities provide the same information in many modalities (e.g., visual alert plus audio warning) to increase resilience. Encourage a range of user skills by providing options for those with disabilities, such as hearing loss or mobility limitations. Personalization permits input/output preferences to be changed.

Adjust for context; determine if the user is in a crowded, noisy, or moving environment and modify the modes of interaction appropriately. For dynamic modality selection, this function automatically selects or recommends the optimal interaction mode. Keep the user from being overloaded; having too many modalities open at once can be annoying. Regarding explicit comments, make sure users are aware of the status of the system and how it reacts to their inputs. Ensure that the system correctly synchronizes inputs and outputs, especially when combining gesture and speech or vision and touch. In dealing with latency, minimize the duration needed to process or integrate information from multiple modalities. In graceful degradation, the system should revert to other options in the event that one modality malfunctions (for example, the microphone is blocked). Multimodal error correction allows users to quickly address misunderstandings using a different modality [11]. If processing sensitive data, extra protection may be needed for voice or physiological data. When collecting data from cameras, microphones, or other sensors, users should be clearly informed to obtain their consent. If multimodal inputs yield consistent results across contexts, then the responses are consistent. In a unified user interface, ensure a consistent flow of interactions across all modalities.

(iv) Applications and Real-World Examples of Multimodal Interaction

In autonomous and semiautonomous vehicles, gesture controls such as BMW's gesture control are all inputs utilized in multimodal interfaces for driver assistance systems. Voice alarms, tactile steering feedback, and HUDs are examples of outputs. The objective is to reduce distraction, promote natural interaction, and ensure safety. In industrial robotics/smart manufacturing (Industry 4.0), voice control, wearable sensors, and gesture recognition such as waving to halt a robot are all inputs utilized in human–robot collaboration (HRC) systems. Collaborative robots use haptic feedback, aural alarms, and visual cues as outputs. Robots such as KUKA and ABB, which combine speech and gesture control for versatile assembly tasks, are a couple of examples.

In healthcare and assistive technologies, eye tracking, BCIs, remote monitoring, and voice input for those with limited mobility are all inputs used in rehabilitation systems. The outputs include robotic limb actuation, screen instructions, and audio feedback. Smart wheelchairs and prosthetics that integrate speech, joystick, and EMG input are a few examples. In smart homes/ambient assisted living, voice (Alexa, Google Assistant), gesture (*via* cameras), and touch panels are used as input methods in voice plus sensor-based interaction. The outputs are screen displays, spoken answers, and illumination indications. A few instances include, to guarantee proper room reaction, the system combining the user's words "turn off the lights" and their pointing.

In gaming and virtual/AR, voice, motion tracking, eye movement, and gesture (VR gloves) are all inputs utilized in immersive multimodal interaction. The outputs include visual immersion, 3D audio, and haptic feedback. The Meta Quest and PlayStation VR, which use gesture plus speech plus controller inputs, are two examples. In aviation and mission-critical systems, touchscreens, voice instructions, and eye tracking are all input methods utilized in cockpit assistance systems. The outputs include force-feedback, graphic panels, and audio alerts. Military pilot helmets with integrated HUDs, voice communications, and gaze tracking for target selection are a few examples. In public spaces and transportation, inputs for multimodal kiosks and navigation systems include touch, speech, and gesture (for accessibility). The output consists of audio and visual instructions. A few instances are voice-activated and touch-activated smart kiosks found in malls and airports. In construction and field work, voice, hand motions, and gaze tracking are all used as input methods in wearable AR systems. On smart glasses (such as the Microsoft HoloLens), audio instructions are produced along with visual overlays. A few examples are technicians using AR plus speech plus gesture in complicated locations to retrieve manuals hands-free.

(v) Benefits of Multimodal Interaction in CPHS

By simulating human communication (such as speech and gestures), systems become simpler and easier to use. Users have a lower learning curve because they are accustomed to the interaction patterns. Systems can function even in the event that one modality fails thanks to redundant inputs (e.g., fallback to gesture if speech fails). Having more avenues for clarification or correction facilitates error recovery. Inclusiveness and accessibility features help people with impairments by providing alternate input/output modalities (e.g., touch for the hearing impaired, speech for the vision impaired). Users are able to select or mix their favorite modality. Systems are able to change their behavior in response to their surroundings. For example, they can transition from voice to text in noisy environments. System intelligence is increased through richer context inference made possible by multimodal input.

Quicker and more effective interaction results because the execution of tasks is sped up by the simultaneous use of modalities (e.g., voice plus gesture), reducing the necessity for sequential involvement, especially in time-sensitive scenarios. Immersion and interaction boost user satisfaction, particularly in VR/AR, education, and gaming. Interaction becomes more enlightening and fulfilling when multimodal feedback (visual, aural, and tactile) is used. In high-risk situations, it permits hands-free or eyes-free operation (e.g., voice commands for pilots or surgeons). The likelihood that a user will notice important warnings is increased by multimodal alerts (visual, auditory, and haptic). Multimodal orchestration enables users to carry out complicated tasks, such as using voice commands, gestures to point, and eye gaze to select, and it enhances CPHS performance in dynamic, uncertain settings.

3.3.2 Sensor Integration in Advanced HMI for CPHS

Human–machine interaction is a crucial link between users and intricately linked physical–digital environments in CPHSs. The integration of various sensors into HMI systems becomes crucial as CPS across industries such as manufacturing, healthcare, transportation, and smart infrastructure grow in complexity and autonomy. Multimodal, real-time, and context-aware interaction is made possible by sensor integration, which enables machines to understand human intents, decipher environmental cues, and react intelligently. HMIs are evolving from passive input devices to intelligent, active interfaces that improve efficiency, safety, and usability.

(i) Types of Sensors Used in HMI for CPHS

The few types of sensors used in HMI for CPS are vision-based sensors, audio sensors, touch and force sensors, motion and position sensors, physiological and biometric sensors, and environmental sensors. A family of sensors known as vision-based sensors uses cameras or other optical devices to gather and process visual data from their surroundings. Enabling machines to see, understand, and react to their environment similarly to human vision is the main objective of vision-based sensing. Activities such as motion tracking, object detection, depth estimation, and scene comprehension fall under this category. Audio sensors can pick up and record sound waves from their environment, converting them into electrical signals for processing and analysis. Since audio sensing enables robots to perceive, understand, and engage with their auditory environment, it is an essential component of many modern technologies. Touch and force sensors are vital components of systems that require physical interaction with the environment. Because these sensors enable the detection of contact pressure or force applied to a surface, they are essential for tactile feedback, precise control, and human–machine interactions. Touch and force sensing must be combined to build complex haptic systems that allow machines to both perceive and respond to tactile feedback.

Motion sensors detect motion and changes in velocity or direction. These sensors allow systems to maintain stability, understand human gestures, and track movement trajectories in real time. On the other hand, position sensors establish the precise location of an object within a given space. They are crucial for a variety of applications such as AR devices, robotic arm control, industrial machinery, and navigation systems. Physiological sensors are used to measure internal body signals such as brain activity (EEG) and oxygen saturation (SpO_2). They provide real-time information about a person's health status, which helps with early abnormality diagnosis and continuous health evaluation outside of traditional clinical settings. In contrast, biometric sensors focus on identifying individuals based on their distinct biological or behavioral traits including their voice, gait, fingerprints, iris patterns, and facial features. Smartphones, biometric passports, access control, and other security and authentication systems commonly use these sensors. Environmental sensors are devices designed to measure, detect, and monitor the physical and chemical conditions in the immediate environment. These sensors provide essential environmental information including temperature, humidity, air quality, light intensity, gas concentration, sound levels, radiation, and atmospheric pressure. As sustainability, health, and climate resilience gain more attention globally, environmental sensors are becoming essential tools for real-time environmental evaluation and decision-making.

Environmental sensors continue advancements in sensitivity, reliability, and affordability as technology advances drive innovation in climate monitoring, environmental preservation, and sustainable development.

(ii) Role of Sensors in Enhancing HMI Capabilities

Because they allow machines to receive, understand, and react to human input more naturally and intelligently, sensors are essential to the advancement of HMI technology. Sensors offer the vital connection between the physical and digital worlds, enabling smooth human–machine interaction as the need for intuitive, responsive, and context-aware systems increases. In applications such as smart displays, car entertainment systems, and assistive technologies, vision-based sensors enable touchless control and customized responses by enabling machines to understand gestures, facial expressions, and user presence. Voice recognition and sound-based interactions are made possible by audio sensors, which are crucial for hands-free communication, voice-activated assistants, and accessible solutions for people with impairments.

Touch and force sensors enhance accuracy and responsiveness in robots, prosthetic devices, and touchscreens by providing tactile feedback and pressure-sensitive controls. In wearables, AR/VR settings, and interactive gaming, motion and position sensors facilitate gesture-based user interfaces, movement tracking, and adaptive system responses. By identifying people and modifying system behavior in response to biometric identification or real-time health data, physiological and biometric sensors enhance security and user-centric functionality while personalizing the HMI experience. By enabling systems to adjust in response to external factors such as temperature, illumination, or air quality, environmental sensors improve context awareness and facilitate more intelligent decision-making in residential, commercial, and automotive contexts. Modern HMIs are able to predict actions, comprehend complicated user intent, and deliver adaptive feedback by utilizing ML and combining multisensor data. HMIs are becoming more accessible, immersive, and natural as a result of the transition from simple input techniques to intelligent, multimodal interfaces.

(iii) Benefits of Sensor Integration in HMI

Adding several sensors significantly enhances the usability, responsiveness, and user experience of HMIs. The primary benefits include improved user experience, interaction in several modes, customization, context awareness, and better security and safety. Sensors provide more natural and intuitive interactions through touch, gesture, speech, and presence detection, making system use easier and more enjoyable. By integrating inputs from several sensors (such as speech, motion, and facial recognition), users can interact in a number of ways, increasing accessibility and flexibility.

The ability of biometric and physiological sensors to tailor system behavior according to individual characteristics, preferences, or emotions (e.g., identity or stress level) may lead to more relevant and adaptive interactions. Motion and environmental sensors enable systems to comprehend situational context including location, movement, and ambient conditions and modify behavior appropriately (e.g., reducing noise in quiet environments or dimming screens in low light). Sensors enhance both operational and personal safety by monitoring dangerous situations (such as gas leaks and excessive force) and facilitating user authentication (such as fingerprint and face identification).

3.3.3 AR and VR in Advanced HMI for CPHS

CPHSs are composed of computational algorithms and physical processes connected by communication networks. These systems are becoming increasingly complex in autonomous vehicles. Advanced HMIs are necessary for managing, monitoring, and interacting with CPHS in an efficient manner. The natural, immersive, and context-rich interactions needed for complex CPS contexts are frequently not provided by traditional interfaces. By reducing the distance between people and cyberphysical environments, AR and VR have become game-changing technologies that improve human–machine interaction. AR and VR let users better understand, control, and communicate inside CPHS through immersive visualization and natural interaction modalities, which improves operational efficiency, safety, and decision-making.

Through the application of AR, users can see real-time sensor data, system statuses, and alarms contextualized within their physical surroundings. Users can explore and engage with virtual representations of CPHS components and procedures in completely immersive computer-generated worlds offered by VR [12]. These skills are essential in fields where operational understanding is enhanced by geographical awareness and real-time data integration, such as energy systems, smart cities, and industrial automation. AR/VR interfaces use haptic feedback, eye tracking, voice instructions, and hand gestures to provide natural interactions. Users can operate complex systems more effectively and intuitively by substituting interactive 3D representations for conventional 2D screens and controls. AR/VR solutions, when combined with digital twins (DTs) and real-time sensor data, allow operators to remotely monitor CPHS, do diagnostics, and work with others from any place, improving responsiveness and decreasing downtime [13]. VR provides secure, realistic training environments that mimic actual CPHS situations, enabling users to rehearse protocols, solve issues, and get ready for crises without endangering people or property. By adapting to various user roles and preferences, AR/VR-driven HMIs can increase accessibility for a wide range of operators and improve usability across skill levels.

(i) Key Technologies Behind AR/VR Integration

A number of key technologies are required for the effective integration of VR and AR into HMIs for CPSs. Real-time data visualization, immersive interactivity, and smooth synchronization between virtual and actual surroundings are all made possible by these technologies. In head-mounted displays, by projecting 3D content into the user's field of vision, devices such as the HTC Vive, Meta Quest, and Microsoft HoloLens offer immersive AR/VR experiences. Smart glasses are thin, transparent AR screens that project data onto the physical world while being mindful of the environment. The projection-based AR technique, which is helpful in collaborative or industrial contexts, overlays digital content onto real-world surfaces using external projectors. Motion tracking sensors include optical tracking systems, magnetometers that detect movement of the head, hands, and body, and IMUs (accelerometers, gyroscopes). LiDAR and depth sensors provide depth perception and spatial mapping, which are essential for lining up virtual objects with actual geometry. Eye and gesture tracking can be done by keeping track of the user's hand and eye motions; this feature enables more organic engagement.

Simultaneous localization and mapping (SLAM) and spatial mapping algorithms are essential for AR/VR systems to understand and track the user's surroundings in real time. They ensure stable interactions among users and sessions and help virtual components effectively integrate into the physical space. Realistic virtual worlds and interactive simulations are made with 3D graphics engines such as Unity and Unreal Engines. These engines offer a realistic and responsive user experience by supporting physics modeling, lighting, texturing, and real-time feedback. AR/VR systems can remotely receive real-time data from CPS and offload processing thanks to edge and cloud computing. For seamless, synchronized experiences, particularly in distant or remote applications, 5G and low-latency networks are essential.

A virtual version of a physical system that is updated constantly with data from the real world is called a DT. Users can see, track, and engage with live CPS activities in an immersive setting by combining AR/VR with DTs. Identifying motions, users, and items and forecasting how users will behave can be done using adaptive interfaces. Enhancing spatial awareness and SLAM makes it possible to use AR/VR interfaces to engage with CPS in a more sophisticated and customized way. These gadgets create virtual worlds by simulating touch. Gloves, vests, and handheld controllers that offer force feedback are a few examples that enhance realism and interaction.

(ii) Applications of AR/VR in Cyberphysical Systems

CPSs and AR/VR technologies have combined to create new opportunities for interactive simulation, improved visualization, and intuitive control. Real-time viewing of machine condition, process flow, and maintenance instructions superimposed on actual equipment is made possible by AR in smart manufacturing (Industry 4.0). VR is used to simulate production lines for layout optimization and to teach operators in intricate assembly processes. Benefits include less downtime, accelerated onboarding, and increased operational effectiveness. AR improves the precision of image-guided surgery by superimposing computed tomography or magnetic resonance imaging scans on the patient's body. VR is used in surgical simulation training, therapy, and rehabilitation. Benefits include safer training settings, better patient outcomes, and increased surgical accuracy. AR makes it possible for city planners and maintenance teams to see subterranean infrastructure, including electrical grids and water lines, in smart cities. VR is used to model urban settings for infrastructure design, traffic flow optimization, and disaster preparedness. Benefits include improved public safety, lower maintenance costs, and better planning choices.

AR helps maintenance staff by providing overlay instructions for complicated systems in spacecraft or airplanes. Mission simulations, pilot training, and combat visualization are all done *via* VR. Benefits include reduced operational risks, better mission preparedness, and improved training. In power plants and utility grids, AR enables field technicians to see electrical circuits, sensor data, or pipe networks. Control rooms and dangerous situations (such as nuclear incident drills) can be vividly simulated with VR. Benefits include increased system reliability, quicker troubleshooting, and safer field operations. AR in cars gives heads-up displays, contextual information, navigation, and notifications. Driver behavior analysis and autonomous driving simulations both make use of VR. Benefits include improved user experience in car interfaces, quicker R&D, and increased driving safety. AR/VR systems are utilized to teach complicated STEM (science, technology, engineering, and mathematics) topics, replicate lab environments, and offer hands-on instruction in safe situations. Benefits include scalable technical instruction, safer learning, and higher levels of engagement.

(iii) Benefits of AR/VR in HMI Design for CPHS

Human interaction with CPSs is greatly improved by the incorporation of AR and VR into HMI design. A more natural, effective, and educational interface layer is provided by these immersive technologies. Through the use of gestures, gaze, or voice instructions, AR/VR allows 3D spatial interaction, improving interface naturalness and lowering the cognitive load associated with interpreting system data. Operator training time is decreased and learning curves are accelerated. Live data, alarms, and system conditions are visualized using AR/VR and superimposed on real-world devices or in realistic simulations. This makes it easier for users to quickly understand the state of the system. Benefits include improved operational safety and speedier decision-making.

Users can see inside machines or comprehend system dynamics in real time by using AR or VR to visualize complex system operations and sensor outputs in three dimensions. Benefits include improved process optimization, fault identification, and diagnostics. Through AR/VR interfaces, users may access and manage CPHS remotely, facilitating real-time communication between specialists, engineers, and field operators in various places. Faster problem solving and less downtime are achieved. Compared to conventional 2D dashboards, AR/VR interfaces can simplify complex data and decrease clutter by utilizing 3D space and spatial memory. This increases user satisfaction and usability. Depending on the role, behavior, or physiological information of the user (e.g., fatigue level from biometric sensors), AR/VR systems can modify the content and control options. This leads to improved user performance, comfort, and focus. By safely simulating real-world situations, VR offers risk-free environments for teaching people on CPHS operation, maintenance, and emergency protocols. This increases readiness and retention of skills.

3.4 Enabling Technologies for Advanced HMI in CPS

The design of HMIs must change as CPS becomes more dynamic, sophisticated, and data-intensive to guarantee efficient, secure, and user-friendly human–machine interaction. The needs of contemporary CPS in industries such as smart manufacturing, healthcare, energy, and autonomous systems cannot be met by traditional HMIs, which are frequently restricted to simple graphical interfaces and manual controls. A number of cutting-edge technologies are being incorporated into HMI deployment and design to meet this challenge. The foundation of next-generation HMIs is made up of these enabling technologies, which include secure communication protocols, DTs, EC, AI, ML, and interoperability standards.

3.4.1 AI and ML in Advanced HMI for CPS

HMIs must change from static control panels to dynamic, intelligent interaction systems as CPHSs grow more autonomous and data-rich. ML and AI are at the center of this shift. They enable HMIs to react to user behavior, learn from data, and make tenable decisions in real time. Many industries such as manufacturing, transportation, healthcare, and energy can greatly improve performance, safety, and efficiency by incorporating AI and ML capabilities into HMIs to create user experiences that are more resilient, intuitive, predictive, and personalized.

(i) Applications of AI and ML in Advanced HMI for CPS

AI-powered HMIs can forecast user needs based on historical usage patterns and system conditions. An example would be an industrial control interface that points out the maintenance procedures that must be followed before a failure occurs. ML algorithms adapt the design, functionality, or complexity of the HMI dynamically based on user experience, context, or workload. For example, a smart factory's interface provides advanced insights to experts while simplifying tasks for novice operators. AI enables chatbots that allow users to interact with CPHS in natural language and voice-activated instructions. An example would be a voice assistant that can be linked to a building management system and respond to spoken commands to adjust the lights and HVAC.

To detect unusual activity and potential system issues, ML models analyze real-time data. To encourage proactive decision-making, the HMI can communicate these insights. An example would be a power grid control interface that uses ML predictions to notify operators of possible transformer faults. AI can modify interaction styles and content delivery by deciphering user emotions, movements, and biometric inputs. An example would be an HMI for driving assistance that recognizes driver weariness and reacts by recommending autopilot or issuing alerts. AI-powered HMIs facilitate complex decision-making by offering context-aware visuals, automatic control alternatives, and real-time recommendations. An illustration would be a surgical interface that makes recommendations for steps to take based on continuous evaluation of patient information and surgeon activity.

(ii) Benefits of AI and ML in CPS

The HMI is customized by AI and ML algorithms according to user roles, preferences, behaviors, and habits. This includes improved user satisfaction, less cognitive strain, and enhanced comfort. ML models predict user behavior, maintenance requirements, or system behavior before they happen. This promotes proactive decision-making, enhances planning, and reduces downtime. Depending on the user's level of experience or the situation at hand, interfaces dynamically modify their layout, information density, or control options. This leads to increased productivity and fewer mistakes due to better usability for both inexperienced and seasoned users. AI and ML analyze real-time data streams to find anomalies or breakdowns in system behavior.

This facilitates prompt action, increasing operational safety and reliability. More intuitive methods of interacting with systems are made possible by the integration of voice control, gesture detection, and NLP. This allows for more effective and accessible communication, particularly in high-risk or hands-busy settings. Based on real-time system analysis, AI-enhanced HMIs can offer recommendations, automated responses, or actionable insights. This lessens operator strain and facilitates precise, data-driven choices in urgent situations. Over time, ML models can enhance interface behavior by learning from operational data, user feedback, and environmental changes. As interfaces develop, they become more efficient, sensitive to context, and user-friendly. With little modification, AI and ML-powered HMIs can be used in a variety of CPHS applications, such as healthcare systems and smart factories. This lowers development costs and boosts system adaptability.

3.4.2 Digital Twin Technology in Advanced HMI for CPS

A DT is a real-time, virtual model that replicates the behavior, state, and functionality of a physical system, procedure, or asset. DTs are essential for connecting the digital and physical realms in CPSs so that control, monitoring, and interaction can occur smoothly. DT technology turns conventional interfaces into intelligent, interactive, and immersive platforms when combined with HMIs. Through interactive simulations, predictive insights, and dynamic representations, it improves the user's comprehension of complicated systems. DTs are becoming more and more crucial as CPS continue to expand in size and complexity to guarantee that human operators can communicate with them effectively and naturally.

(i) Role of Digital Twin Technology in CPS and HMI

The digital and physical layers of CPS are connected in real time by DT technology. By combining information from sensors, control systems, and analytics platforms, it continuously replicates the condition, actions, and output of physical systems. System modeling, which digitally represents physical assets and processes for analysis and control, is one of the primary functions in CPS. The real-time synchronization method uses real-time sensor and actuator data to update the digital model. Operational optimization makes use of analytics and simulations to find inefficiencies and provide fixes. Diagnostics and fault detection support proactive maintenance by identifying irregularities and forecasting problems. Design and testing allow for the testing of system modifications and virtual prototyping without interfering with actual operations.

DTs offer users intelligent, interactive, and visual interfaces for monitoring, comprehending, and controlling CPS in the context of HMI. One of the primary functions is interactive visualization, which improves situational awareness by displaying a digital representation of actual assets in real time. Decision support provides simulation findings and prediction insights to help operators make decisions. User training offers operators realistic practice settings free from real-world dangers. Regarding remote control and monitoring, using twin models connected to the cloud, users can remotely operate CPS. Adaptive interfaces modify the information shown according to the role, requirements, or circumstances of the user.

(ii) Applications Across CPS Domains

The development of data-driven, real-time virtual copies of physical systems is made possible by DT technology. DTs offer a vital basis for enhancing the monitoring, control, simulation, and interface capabilities of CPSs. A thorough examination of the ways in which DTs are transforming different CPS industries is provided below, broken down by domain. In the smart manufacturing DT use case for HMIs, operators use a real-time 3D DT of the manufacturing line to monitor productivity, identify delays, and simulate changes to the production schedule. Advantages consist of reduced downtime, enhanced decision-making, and faster responses to machine issues.

For real-time intervention, patient monitoring systems give doctors a DT of the patient's vitals, organs, or implants. Benefits include faster critical care response times and more accurate diagnoses. Real-time twins of autonomous car fleets are used to monitor traffic patterns, vehicle health, and system irregularities, with vehicle control. Benefits include predictive maintenance, remote operation, and safe fleet management. Control room dashboards give operators real-time decision-making tools, model failure scenarios, and show power flow. Benefits include increased grid dependability and quicker disturbance recovery. To maximize energy efficiency and comfort, facility managers communicate with a building's digital counterpart to regulate the HVAC, lighting, and security systems. Benefits include increased occupant safety, proactive maintenance, and energy savings.

(iii) Benefits of Digital Twins in HMI for CPS

By offering real-time, intelligent, and context-aware interfaces, DTs integrated into HMIs greatly improve human interaction with, monitoring of, and control of CPSs. Using 2D/3D interfaces to visualize system states in real time allows operators to see inside the system to comprehend dynamics, behavior, and flaws. A wind turbine's DT displays real-time temperature, vibration, and stress data. Operators can investigate what-if situations through simulation-based forecasting. This provides early alerts for abnormalities, malfunctions, or system deterioration. Before a motor breaks, a factory predictive dashboard notifies users.

HMI changes according to context and user role (e.g., engineer versus operator). By giving priority to the most pertinent information, interfaces can lessen cognitive strain. During surgery, surgeons receive streamlined real-time metrics, and engineers view comprehensive postoperative diagnostics. DTs make it possible to test and simulate without endangering real systems. This encourages safe experimentation, particularly in vital systems such as electricity grids and airplanes. Before implementing load changes in real life, operators model them on a digital grid twin. This allows for remote collaboration, control, and system visualization, which is vital for dispersed systems (such as distant factories or offshore platforms). From a different city, a remote technician diagnoses and configures equipment using a DT.

Predictive maintenance tools are included into DT-based HMIs. Unplanned failures are reduced through condition-based service and scheduling. A robotic arm's digital doppelganger advises maintenance schedules and calculates the arm's remaining useful life. DTs combined with immersive HMI offer training simulations that are realistic. This lowers the learning curve for sophisticated machinery. Before working with the actual setup, trainees engage with a virtual refinery system. The smooth combination of user interaction, AI analytics, and sensor data allows for continuous system optimization made possible by DTHMI loops. Irrigation systems in smart agriculture automatically modify their settings in response to farmer input and twin feedback. Because they are interactive and visual, users can better comprehend the intricate choices made by autonomous CPS with the aid of DTHMIs. This enables explainability, which promotes confidence in automation. The HMI in autonomous cars allows drivers to see the reasoning behind decisions.

(iv) Digital Twin Architecture in CPHS

Buildings, infrastructure, or systems can be virtually represented through the use of DT architecture. This digital version mimics the traits, behaviors, and functionality of actual resources. Real-time data and advanced modeling approaches make it possible. Therefore, it is more crucial to integrate real-time data in DTs. The use of DT software is growing quickly in the building and architecture sectors. From planning and development to maintenance and operation, DTs give architects, designers, and project managers the ability to view, evaluate, and optimize each phase of a building's life cycle. The way that DTs function is by producing a computer image that closely resembles actual objects. This DT platform facilitates stakeholder interaction, assessment, and optimization of building and system lifecycle processes. Figure 3.3 shows the blocks of the DT in CPHS.

Sensors, IoT devices, and historical documents are the sources of data. Real-world items, systems, settings, and performance are all covered by this data. Sensors, for instance, monitor temperature, humidity, energy use, and structural integrity in real time. A 3D digital model of the building is produced using the collected data. This simulation is dynamic and adapts to new data input; it is not static data. By ensuring accuracy and realism, the majority of contemporary platforms and software tools aid in the development of these models. Real-world processes are synchronized in real time by DTs. The digital model instantly reflects any modifications. To determine how the buildings function in various scenarios, DTs execute a number of simulations. This comes after environmental impact assessments, energy consumption production, and stress tests. These virtual simulations and models identify potential hazards and optimize the building's design and operations. Creating a continuous feedback loop, wherein DTs receive input from the real world, is the final phase. This ongoing

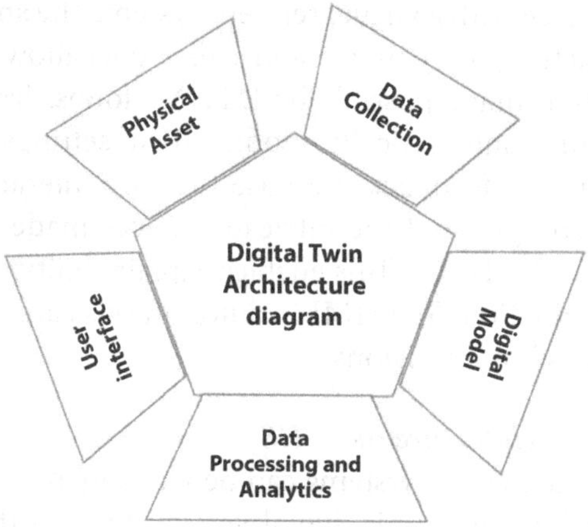

Figure 3.3 Digital twin flow diagram in CPHS.

procedure guarantees that the digital model is accurate and up to date, providing insightful information for upkeep and operational efficiency.

(v) Advantages of Digital Twin Architecture

Here are a few benefits of the DT architecture. Figure 3.4 shows the advantages of DT architecture. By enabling accurate energy modeling, material improvement, better construction, and waste reduction, DT architecture contributes to the sustainability of building projects. As a result, DTs are used in sustainable building to simulate different situations. These aid project teams in identifying the most practical and environmentally responsible ways to mitigate the environmental effects of their projects.

Applications for DTs enhance planning and visualization by offering an engaging and interactive 3-D model. As a result, the stakeholders can find a variety of design possibilities, spot mistakes early, and make better judgments that support the project's objectives. This lessens mistakes and the need for costly solutions to the problems. As a result, DTs increase productivity while cutting expenses and time. In the construction industry, DTs aid in providing up-to-date data on the performance of a building or structure over the course of its lifetime. Sensors that contain resources from the real world are used to enter data into the DT model. As a result, this model can provide information about environmental conditions, energy usage, and structural integrity. By increasing the buildings' lifespan and effectiveness, this ongoing maintenance and feedback enables proactive maintenance.

The architecture of DT platforms facilitates better collaboration and communication between project managers and team members. The

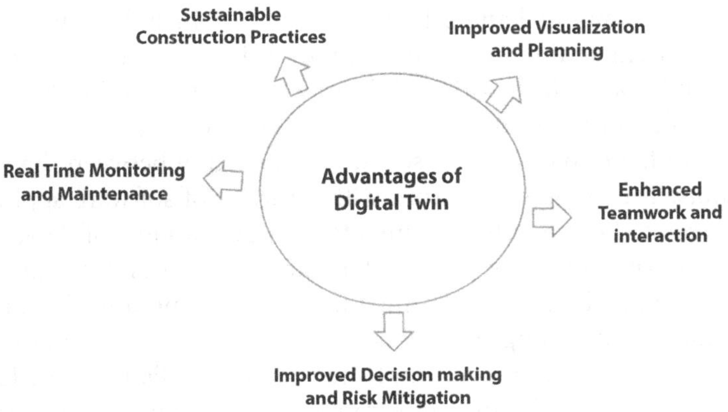

Figure 3.4 Advantages of digital twin architecture.

architects, engineers, contractors, and other shareholders can collaborate more efficiently thanks to the DT models, which aid in creating an updated and shared virtual representation. Teamwork is enhanced by the ease with which any irregularities can be shared. This transparency promotes better project completion and boosts productivity with DT solutions. The wealth of data and predictive analytics available to decision-makers helps them make better decisions. Thorough design testing and validation can also benefit from these analytics prior to construction. Thus, by identifying and resolving the possible risks and hazards in building projects, the project team can avoid issues and costly maintenance before construction begins.

(vi) Challenges

To use DTs effectively, businesses need to overcome several challenges when integrating them into architecture. Creating a DT model is one of the hardest tasks because it involves combining data from multiple sources. A lot of systems, sensors, and platforms are typically involved in architecture projects; these components need to be precisely synchronized for the DT to work properly. Ensuring data quality is crucial, because inaccurate or missing data might result in a flawed DT model, which can affect the model's dependability and the decision-making process. DTs have many long-term benefits, but they can come with hefty initial expenses for technology, infrastructure, and talent development. This includes the cost of the necessary hardware, software, and expertise to set up and manage DT systems. Producing a realistic DT model that faithfully captures the asset, structure, or architectural system may prove difficult. Because of the model's level of detail, advanced technologies and expertise are required, particularly for intricate and extensive projects.

As DT technology advances, companies may run into issues expanding their solutions to handle larger projects or include new technologies. DTs for expanding infrastructures or projects may be challenging to manage and update if the original setup was not designed to be scalable. Because different platforms and software may not always be compatible, it may be difficult to establish a seamless connection between the DT and other tools. The DT needs to work with a variety of software applications and systems to be properly implemented. Large amounts of data, such as building information, operational details, and sensitive architectural drawings, are handled by DTs. For companies that incorporate DTs into their architecture, maintaining privacy compliance and safeguarding the data from cyberattacks are crucial. Implementing DTs calls for knowledge of ML, data analytics, and system integration. To properly set up and manage these systems, businesses must invest in recruiting or training people.

3.4.3 Edge Computing and Cloud Integration in Advanced HMI for CPS

Enhancing the functionality, responsiveness, and scalability of sophisticated HMIs in CPS requires the combination of EC and cloud services. By striking a balance between robust, centralized data processing capabilities and real-time responsiveness, these technologies enhance one another. EC makes interactions smarter and more adaptive in advanced HMI for CPS by enabling real-time response, low latency, and improved dependability. Local data processing close to the data source, usually at or close to the sensors, actuators, or HMI devices, is made possible by EC. For safety, control, and user engagement in real-time CPS applications, this close proximity guarantees instantaneous response times and drastically lowers latency. Edge nodes, for instance, locally process sensor data in autonomous cars or industrial automation to allow for immediate decision-making without the need for distant servers.

Conversely, cloud integration offers centralized resources for long-term system learning, deep analytics, and massive data storage. The cloud aggregates data from several edge devices in various places, enabling predictive maintenance, user behavior analysis, and system improvement. By shifting computationally demanding operations to the cloud, edge nodes and HMIs can continue to be small and energy efficient. Hybrid HMI models are made possible by the synergy between edge and cloud architectures, where complicated data processing, AI model training, and historical data visualization are carried out in the cloud, whereas time-sensitive interactions are managed at the edge. These structures provide individualized interaction patterns in CPS, adaptable user interfaces, and smooth user experiences across devices.

(i) Key Benefits of Integrating Edge and Cloud in Advanced HMI for CPHS

Instantaneous data processing near the source is made possible by EC, which guarantees quick system response, which is necessary for real-time feedback, control, and user interaction. Because cloud infrastructure offers nearly limitless computing and storage capacity, the CPHS can effectively increase with the volume of data or complexity of the system. In time-sensitive applications, edge nodes can continue to perform vital functions even in the event of connectivity outages, guaranteeing system availability and local autonomy. Only necessary or aggregated data is sent to the cloud after local processing and filtering, which lowers network traffic and boosts productivity.

Complex analytics, processing of historical data, and training of ML models are all supported by the cloud and may be delivered back to the edge for inference in real time. The user experience is enhanced by dynamic HMI customization depending on user preferences, behavior, or environmental context made possible by data-driven insights from the cloud. Although anonymized, aggregated datasets can be handled in the cloud for more extensive analysis, sensitive data can be processed at the edge, reducing the risk of exposure. Cloud systems make it easier to remotely manage, diagnose, update, and coordinate dispersed edge devices and HMI elements. By shifting complex calculations to the cloud, the processing load of edge devices is lessened, resulting in reduced battery usage and longer device life. The HMI is adaptable and interoperable in intricate CPHS ecosystems thanks to edge-cloud connectivity, which allows for smooth operation across a variety of platforms and devices.

(ii) Hybrid Edge-Cloud HMI Architecture

By processing data locally, edge devices (such as embedded HMIs, programmable logic controllers, or smart sensors) enable real-time decision-making and lessen reliance on centralized cloud systems. To communicate directly with other devices and cloud platforms, these devices can also host microservices and APIs. Complex analytics, long-term storage, and wider

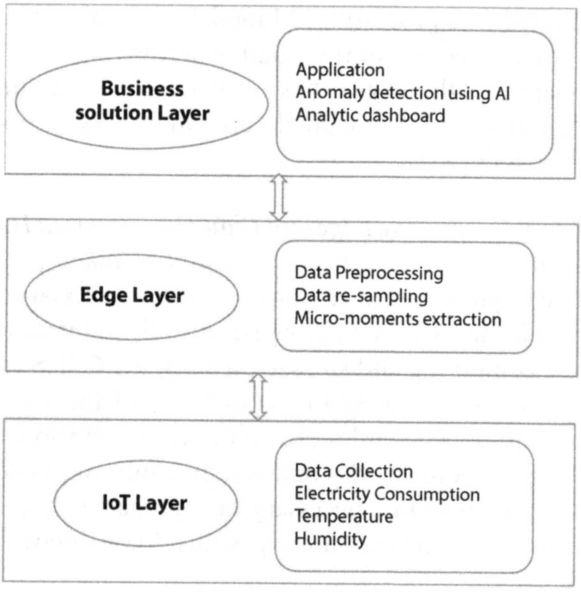

Figure 3.5 Hybrid edge-cloud HMI architecture.

system interaction are managed by the cloud layer. Figure 3.5 shows the outline of hybrid edge-cloud HMI architecture.

(iii) Role of the IoT Layer

A variety of smart devices, including sensors, actuators, and controllers, are part of the IoT layer. These devices record data, such as temperature, humidity, and motion, and carry out commands, such as turning on lights and modifying thermostats. This layer uses protocols such as cellular networks to guarantee connectivity between devices and higher levels (cloud or edge). Prior to sending information to the cloud, some IoT devices can evaluate data locally and make decisions using EC capabilities [14].

(iv) Role of the Edge Layer

By processing data near its source, the edge layer lowers latency and bandwidth consumption. This is especially important for time-sensitive applications that need to make decisions right away. It does preliminary analysis, filters out noise, and aggregates data from multiple IoT devices. By ensuring that only pertinent and processed data is transferred to the cloud, bandwidth and storage are optimized. The edge layer facilitates real-time device control and monitoring, allowing for prompt responses to changes in the environment or system conditions [15]. The edge layer enhances security and helps with regulatory compliance by limiting data transmission to external servers and processing sensitive data locally.

(v) Role of the Cloud Layer

Combining data from multiple edge devices is made possible by the cloud layers' scalable storage options. This consolidated repository supports historical analysis, long-term data retention, and compliance with data governance standards. Strong computational resources are provided by cloud systems for the execution of complex algorithms, ML models, and data analytics. This capability enables anomaly detection, predictive maintenance, and operational process improvement. By allowing operators to view real-time data and control systems from any location, the cloud layer increases responsiveness and flexibility. This remote accessibility is very helpful for managing scattered assets and activities.

Resources can be dynamically scaled to accommodate fluctuating workloads thanks to cloud infrastructure. The system's elasticity allows it to handle increasing data volumes and processing demands without encountering any performance problems. This integration ensures that operations are consistent across the entire organization.

(vi) Benefits of Edge plus Cloud in HMI

Local data processing guarantees quicker reaction times, which are essential for industrial processes that require quick turnaround times. By processing and filtering data at the edge, less data is sent to the cloud, which saves bandwidth. To reduce cybersecurity concerns, sensitive data can be anonymized and examined locally before being transmitted. EC enables uninterrupted operation even in isolated areas with spotty or nonexistent Internet access. Applications can grow in response to demand without exclusively depending on cloud infrastructure when computing resources are dispersed among a network of smart devices.

3.5 Real-World Applications/Some Use Cases

3.5.1 Industrial Manufacturing

By monitoring the condition of the equipment with sensors and HMIs, CPS can identify problems early and minimize unscheduled downtime. For example, CPS have been used by automakers to improve productivity and optimize processes, resulting in a 30% decrease in downtime and a 20% increase in overall equipment efficiency. On assembly lines, advanced HMIs enable safe and effective HRC, increasing accuracy and productivity. The industry of industrial manufacturing uses procedures including machining, assembling, and fabrication to turn raw materials into completed commodities.

Manufacturers of mining and construction equipment, including bulldozers and excavators, are known as construction machinery manufacturers. Manufacturers of farm machinery, such as harvesters and tractors. Businesses that manufacture tools for the extraction of minerals and other geological resources are known as mining machinery manufacturers. Manufacturers of equipment for the extraction and processing of natural gas and oil are known as oil and gas field machinery manufacturers. In printing machinery, there are manufacturers of bookbinding and printing equipment. Manufacturers of sawmills, woodworking equipment, and paper machines are involved in the processing of wood and paper goods. High levels of expertise and technological innovation define these categories. Since no single company controls a large portion of the market, the business is fragmented, encouraging competition and innovation across a range of specializations.

3.5.2 Occupational Safety and Health

By integrating HMIs with IoT devices, medical professionals may keep an eye on patients' vitals in real time, facilitating prompt interventions and individualized treatment. By giving surgeons real-time overlays of the patient's anatomy, AR-enabled HMIs increase precision during operations. Work-related security and well-being training is crucial for ensuring a secure and sound work environment. It includes teaching employees and bosses about workplace hazards and preventive measures to minimize risks and upgrade efficiency.

The importance of recognizing potential workplace hazards and assessing the associated risks is emphasized in training programs. Implementing effective control strategies is made easier by this proactive approach. Workers must receive the proper training on how to select, use, and maintain personal protective equipment to protect themselves from illnesses and injuries. By teaching employees about ergonomics and safe lifting techniques, the likelihood of musculoskeletal issues can be reduced. To prevent accidents, the curriculum covers fire safety, handling flammable materials, and safe electrical practices.

The International Labour Organization's International Training Centre provides a comprehensive online curriculum that covers important aspects of occupational safety and health (OSH). It consists of 29 modules that concentrate on various aspects of workplace safety including risk assessment, emergency response, ergonomics, and other related topics. Upon finishing the course, students are eligible to get a diploma. The Occupational Safety and Health Administration (OSHA) offers 10- and 30-hour outreach training courses that are taught by OSHA-approved instructors and cover fundamental hazard awareness. Workers in a wide range of sectors benefit from the widespread acceptance of these courses. The American Society of Safety Professionals provides free webinars and materials on subjects such as risk assessment, job safety analysis, and contractor safety management, enabling professionals to keep abreast of best practices.

3.5.3 Smart Home Systems

HMIs in savvy homes empower users to control lighting, heating, and security frameworks through mobile applications, voice assistants, or touchscreen panels. This integration improves user comfort and vitality efficiency. A smart home is one that combines various systems and devices such as appliances, lighting, climate control, and security into a single network that can be controlled remotely using voice assistants, smartphones, or automation programs. Convenience, energy efficiency, and home security are all enhanced by this integration.

Switches and bulbs that can be adjusted and controlled remotely are necessary to create intelligent lighting. Intelligent thermostats maximize energy consumption by learning your preferred temperature and adjusting heating or cooling accordingly. One example of such a device is the Nest thermostat. Cameras, motion detectors, door/window sensors, and smart locks are all parts of smart security systems that can be controlled and watched remotely. Smart appliances include things such as washing machines, ovens, and refrigerators with features such as energy monitoring diagnostics and remote control. As the main hub of control for other devices, voice-activated hubs such as Apple's Siri, Google Assistant, or Amazon Alexa are examples of smart assistants.

3.6 Challenges

CPSs that incorporate advanced HMIs present both significant challenges and revolutionary opportunities. These systems require safe, responsive, and user-friendly interfaces because they integrate real-world operations with computational components. The relationship between human–machine interaction is depicted in Figure 3.6. Below is a summary of the primary challenges and technological factors involved in developing effective HMIs for CPS. A variety of devices and legacy systems that might not be able to communicate with one another natively are regularly integrated by the CPS in the interoperability and integration section. Robust middleware solutions and standardized protocols are required to guarantee smooth communication between these various components. Without international standards, integration could be difficult and development costs could increase. Processing data in real time is essential to enabling timely decision-making because of the massive amount of data generated by CPS. Especially in safety-critical fields such as industrial automation and healthcare monitoring, it is crucial to design HMIs that can efficiently handle and display these data instantly.

Developing HMIs that are both user-friendly and accessible for a broad spectrum of user profiles from highly experienced operators to infrequent users of CPS interfaces is challenging. Interfaces must balance sophisticated functionality and ease of use to serve users with varying skill levels. Concerning security and privacy, CPS is susceptible to cyberattacks due to its interconnectedness. Implementing robust security measures is essential to protecting sensitive data and maintaining system integrity. Additionally,

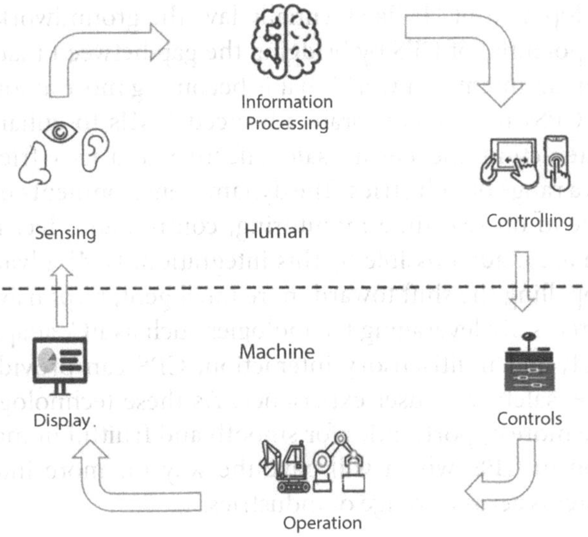

Figure 3.6 Human–machine interaction.

it is crucial to protect user privacy through secure authentication and data anonymization [16]. Users may encounter cognitive overload and modal confusion when utilizing complex HMIs with multiple operational modes. This can lead to errors as they may forget the current system state. Increasing user confidence and safety requires interface design that minimizes mode-dependent behavior and provides unambiguous feedback.

3.7 Conclusion

Enhancing system intelligence, adaptability, and human-centric interaction is made possible by the incorporation of sophisticated HMIs into CPSs. Modern HMI technologies facilitate smooth communication between users and intricate CPS architectures thanks to developments in multimodal interaction, sensor fusion, AR/VR, and ML. In addition to enabling real-time monitoring and control, these interfaces promote personalized experiences and context-aware decision-making. Mission-critical CPS domains such as smart manufacturing, healthcare, autonomous systems, and energy grids depend on low-latency responsiveness and scalable data processing, which are ensured by the integration of EC and the cloud. The constant development of enabling technologies is guiding HMIs toward becoming more intelligent, intuitive, and adaptable despite the difficulties with interoperability, security, and user adaptation.

The development of HMIs essentially lays the groundwork for achieving the full potential of CPS by bridging the gap between machine execution and human intent in a world that is becoming more autonomous and connected. CPSs must incorporate advanced HMIs to enhance human–machine interaction and ensure safe, effective, and user-friendly operations across a range of industries. The dynamic environments in which CPS operate depend on real-time monitoring, control, and decision-making, all of which are made possible by this integration. HMI advancements in CPS are propelling the shift toward more intelligent, responsive, and user-friendly systems. By leveraging technologies such as EC, adaptive control, AR/VR, NLP, and multisensory interaction, CPS can provide enhanced performance, safety, and user experience. As these technologies develop, there will be more opportunities for smooth and fruitful human–machine collaboration in CPS, which will pave the way for more intelligent and effective systems across a range of industries.

AI disclosure statement: The author confirms that no artificial intelligence (AI)-generated content or AI-assisted writing tools were used in the preparation of this chapter (including summaries and the abstract). All text, analysis, and conclusions are the result of the author's original work.

References

1. Dafflon, B., Moalla, N., Ouzrout, Y., The challenges, approaches, and used techniques of CPS for manufacturing in Industry 4.0: a literature review. *J. Adv. Manuf. Technol.*, 113, 2395, 2021.
2. Nandhini, R.S. and Lakshmanan, R., A Review of the Integration of Cyber-Physical System and Internet of Thing. *Int. J. Adv. Comput. Sci. Appl.*, 13, 459, 2022.
3. Habib, M.K., CPS: Role, Characteristics, Architectures and Future Potentials. *3rd Int. Con. on Ind 4.0. and Smart Man.*, 3, 1347, 2022.
4. Radanliev, P., De Roure, D., Van Kleek, M., Santos, O., Ani, U., Artificial intelligence in cyber physical systems. *AI Soc.*, 18, 783, 2021.
5. Tao, F., Qi, Q., Wang, L., Nee, A.Y.C., Digital Twins and Cyber–Physical Systems toward Smart Manufacturing and Industry 4.0: Correlation and Comparison. *Engg.*, 5, 653, 2019.
6. Wang, B., Zheng, P., Yin, Y., Shih, A., Wang, L., Toward human-centric smart manufacturing: A human-cyber-physical systems (HCPS) perspective. *J. Manuf. Syst.*, 63, 471, 2022.

7. Han, B. and Schotten, H.D., Multi-Sensory HMI for Human-Centric Industrial Digital Twins: A 6G Vision of Future Industry. *IEEE Sym. on Comp. and Comm.*, 3, 451, 2022.
8. Krupas, M., Kajati, E., Liu, C., Zolotova, I., Towards a Human-Centric Digital Twin for Human–Machine Collaboration: A Review on Enabling Technologies and Methods. *Sensors*, 24, 2232, 2024.
9. Poursoltan, M., Traore, M.K., Pinède, N., Vallespir, B., A Digital Twin Model Driven Architecture for Cyber-Physical and Human Systems. *Proc. of the I-ESA Conf.*, 12, 135, 2023.
10. Ji, Z., Yanhong, Z., Baicun, W., Jiyuan, Z., Human–Cyber–Physical Systems (HCPSs) in the Context of New-Generation Intelligent Manufacturing. *Engg.*, 5, 624, 2019.
11. Villarreal Lozano, C. and Vijayan, K.K., Literature review on Cyber Physical Systems Design. *10th Conf. on Lear. Fact.*, 10, 295, 2020.
12. Palmieri, M., Bernardeschi, C., Masci, P., A framework for FMI-based co-simulation of human–machine interfaces. *Softw. Syst. Model.*, 19, 601, 2019.
13. Dinter, R., Tekinerdogan, B., Catal, C., References architecture for digital twin-based predictive maintenance systems. *Comput. Ind. Eng.*, 177, 1254, 2023.
14. Liu, Y., Peng, Y., Wang, B., Yao, S., Liu, Z., Review on Cyber-physical Systems. *IEEE/CAA J. Auto. SINICA*, 4, 27, 2017.
15. Sahu, A., Huang, H., Davis, K., Zonouz, S., A Framework for Cyber-Physical Model Creation and Evaluation. *IEEE 20th Int. Conf. on Intell. Sys. App. to Power Sys.*, 20, 134, 2019.
16. Negria, E., Fumagalli, L., Macchia, M., A review of the roles of Digital Twin in CPS-based production systems. *27th Int. Conf. on Flex. Autom. and Intell. Man.*, 27, 939, 2017.

4

Supervisory Control and Data Acquisition Network for Intrusion Detection in Cyberphysical Human Systems

N. Sathish[1]*, P. Arul[2], V. Yokesh[3] and Rajesh Kumar Dhanaraj[4]

[1]Department of Electronics and Communication Engineering, Sri Venkateswara College of Engineering, Sriperumbudur, Tamil Nadu, India
[2]Department of Electronics and Communication Engineering, University College of Engineering, Anna University, Panruti, Tamil Nadu, India
[3]Department of Electronics and Communication Engineering, Sathyabama Institute of Science and Technology, Chennai, Tamil Nadu, India
[4]Symbiosis Institute of Computer Studies and Research (SICSR), Symbiosis International (Deemed University), Pune, Maharashtra, India

Abstract
Supervisory Control and Data Acquisition (SCADA) networks are essential to cyberphysical human systems (CPHSs), enabling real-time monitoring, control, and automation in vital areas such as industrial manufacturing, smart grids, healthcare, and transportation. The growing integration of SCADA with external networks and cloud platforms has rendered these systems vulnerable to various cyber risks, including unauthorized access, virus penetration, denial-of-service assaults, and insider threats. Traditional security measures, such as rule-based firewalls and signature-based intrusion detection systems (IDS), frequently prove insufficient against advanced and developing cyber threats. This chapter examines sophisticated intrusion detection systems tailored for SCADA networks in CPHS. These examine the utilization of artificial intelligence, machine learning, and anomaly detection methodologies to improve threat identification and mitigation. These also explore adaptive security frameworks, behavioral analysis, and blockchain-based security models to guarantee data integrity, confidentiality, and availability. This chapter offers a thorough comprehension of contemporary cybersecurity solutions for protecting SCADA networks from growing security threats.

**Corresponding author*: n.satyish@gmail.com; ORCID: 0009-0000-3507-6337

Anitha Velu, Prasanth Aruchamy, Raghu Ramamoorthy, Rajesh Kumar Dhanaraj and Manjula Arunraj (eds.) Cyber Physical Human Systems: Emerging Trends and Applications, (71–96) © 2026 Scrivener Publishing LLC

Keywords: Intrusion detection systems, automation, data acquisition, block chain, artificial intelligence, machine learning

4.1 Introduction

Cyberphysical human systems (CPHSs) are computing, physical infrastructure, and human intelligence integrated systems aimed to support adaptive and context-aware functionality in areas such as healthcare and smart manufacturing. At the core of such systems is that Supervisory Control and Data Acquisition (SCADA) allows an interface between the digital, physical, and human worlds, which makes on-demand control and monitoring possible. Because SCADA systems are now more connected to one another, they have become highly complex and are at risk of cyber threats. Cyberattacks on SCADA systems have some of the highest profiles, and they illustrate the necessity to implement superior, smart security measures. This chapter explores modern artificial intelligence (AI)–driven intrusion detection methods to safeguard SCADA systems within the evolving landscape of CPHS.

4.1.1 Overview of Cyberphysical Human Systems

CPHSs have marked a paradigm shift in system design in which the computing components, physical infrastructure, and humans are closely combined to realize adaptive, intelligent, and efficient operation. In contrast to traditional cyberphysical systems (CPSs), whose key idea is the interaction between the processes of computation and the physical world, CPHS purposefully embrace the human factors into the control loop as participants and decision-makers [1]. Such synergy between the three aspects enables making decisions in a contextually aware way, human-in-the-loop control, and more naturally satisfying automation systems.

CPHS are now critical to many use cases, such as smart manufacturing, intelligent transportation, digital healthcare, and precision agriculture. Such systems utilize sensor and actuator data in real-time based on AI and machine learning (ML) models, to direct actions and responses that are frequently subject to or reliant on human judgment. CPHS are becoming more refined and larger, and with that, it is more essential that their substructure is reliable, safe, and robust.

4.1.2 Importance of SCADA in CPHS

In numerous CPHS, supervisory control and data acquisition systems provide core constructs to implementations. The SCADA systems have the duty to monitor, control, and collect data on industrial processes as well as infrastructure in real time. They also offer centralized monitoring and control of distributed devices so that the operators can manage and respond to dynamic conditions of complex physical systems.

SCADA systems in smart grids allow the management of electricity flow and the maintenance of grid stability and demand-response operations. Healthcare SCADA systems SCADA-like architectures are used to monitor life-critical life-support systems and facility environments. Under Industry 4.0 conditions, SCADA platforms monitor the activities of robots, logistics, and production lines, and transmit useful information to decision-support systems. The SCADA emerges as an important point in CPHS, connecting the digital, physical, and the human worlds allowing the general intelligence and responsiveness of the system.

4.1.3 Rising Complexity and Connectivity of SCADA Systems

Historically SCADA systems had an isolation going on, and existed in air-gapped models, with a certain level of protection naturally coming because of this. Nevertheless, the trend toward CPHS has brought about the fact that the connectivity and integration of SCADA networks with the enterprise information technology (IT) systems, cloud-based platforms, and even the Internet is required. Such interconnectivity can allow monitoring, predictive analytics, and coordination of the world-spanning infrastructure remotely.

Couched with it is a degree of complexity. Modern SCADA systems consist of heterogeneous devices that use different communication protocols (e.g., Modbus, DNP3, IEC 60870-5-104), and in many cases, such systems do not have in-built security services. A combination of operational technology and IT has boosted the abilities of the system together with its effectiveness to be manipulated by cyberattacks. Furthermore, the system's attack surface is also increased because of human operators and engineers communicating with the system through remote or mobile applications.

4.1.4 Increasing Attack Surface and Real-World Incidents

The increased connection of the SCADA with exterior networks has turned them to be appealing targets of malevolent actors. Intruders use technical vulnerabilities as well as human errors to log into the systems without an authorization, thus paralyzing operations or leaking confidential information. Famous cases of Stuxnet, BlackEnergy, Triton, and attacks on Ukrainian power grid networks have illustrated the potentially devastating breakage of SCADA systems.

In contrast to conventional IT systems, failure of the SCADA systems can cause bodily harm, ecological damage, or death of a human being.

The increased data-driven and autonomous features of the CPHS make them more prone to attack due to specific cyber-related threats due to the exploitation of the interplay between cyberspace and the physical world [2]. These all include denial-of-service (DoS) attacks, man-in-the-middle (MITM) attacks, spoofing, and malware infiltration, and this can be done accurately and with great persistence.

4.2 SCADA Networks in CPHS: Architecture and Functionality

CPHSs rely on SCADAs as central reports, receiving and sending information in real time, such as remote terminal units (RTUs), programmable logic controllers (PLCs), master terminal units (MTUs), and human–machine interfaces (HMIs). Their combination with Internet of Things (IoT), CPS, cloud, platforms, and human frontiers increases the system intelligence and flexibility, along with complexity and security issues [3]. Depending on the trade-offs used, the communication in SCADA networks requires different protocols such as Modbus, distributed network protocol 3 (DNP3), IEC 60870-5-104, and Open Platform Communications—Unified Architecture (OPC UA). Bidirectional data flow is continuous and hence creates situational awareness and responsive control of the cyber, physical, and human levels. The criticality of human operators in the area of decision-making, supervision, and adaptation process strengthens the resilience and the adaptability of SCADA-driven CPHS environments.

4.2.1 Components of a Typical SCADA System

In CPHSs, SCADA systems form the main framework of critical infrastructure and industrial automation. A common structure in a SCADA system is an embodiment of various components that are interrelated in their nature to provide real time monitoring, control, and acquisition. The most important ones are:

(i) **RTUs**: RTUs are microprocessor devices, which can be installed in the field to gather the data of sensors and transfer data to primary systems. They are data concentrators and tend to control actuators including valves, circuit breakers, or pumps.

(ii) **PLCs**: Closely related to RTUs, except that PLCs are normally used in industrial and factory plants, PLCs are programmed forms of logic-based real-time controllers. They are quicker and more flexible and suitable to implement in discrete and process automation.

(iii) **MTU**: The MTU can be described as the processing hub that ensures communication with both RTUs and PLCs. It also gets the sensor data, processes it, and sends commands to the field devices (where appropriate). In more contemporary systems, the MTU is frequently implemented in a control center or on a cloud base.

HMI is a graphical display that enables human operators to see the status of the system, engage in manipulation of data as well as being able to take control when such a need may arise. It converts machine data to human readable formats, offers dashboards, alerts as well as control tools. Completing a SCADA system to combine the structured work of these components allows them to continuously receive and work with environmental or operating information and provides timely and informed decision-making in the cyber, physical, and human domains as well.

4.2.2 Integration with IoT, CPS, Cloud, and Human Interfaces

SCADA systems are getting more integrated and open, making them increasingly networked with information and CPSs in general, the IoT and cloud computing-based services. This embracement serves to make CPHS more functional and scalable and responds faster: the contemporary SCADA systems are rapidly integrating with the IoT devices, which allow them to obtain more real-time and granular data from diverse sources including smart sensors, edge controls, and wearables that constantly update the SCADA servers. SCADA is a vital layer of control that binds the cyber intelligence with the physical infrastructure as the part of the broader CPSs, and cloud platforms further augment these systems by offering high computing power to accomplish tasks such as advanced analytics, predictive maintenance, and AI-based optimization [4].

Moreover, this human-centricity of CPHS systems introduces human actors into a loop by providing modern SCADA interfaces that are compatible with mobile devices, voice interaction, as well as wearable HMIs, including augmented reality glasses, which allows managing a system in a flexible and responsive way, both in regular and emergency situations [5]. Such integration supports situational awareness and adaptability of the systems but it also brings new complexities and security issues especially in regard to real-time performance and points of cyber intrusions.

4.2.3 Communication Protocols in SCADA Systems

The SCADA networks depend on communication. They combine using standard protocols and proprietary protocols in relaying information between field devices and control centers. The most standard protocols are some of the following:

(i) **Modbus (RTU/TCP)**: An open-sourced protocol widely in use since the 1970s for serial communication adjusted to TCP/IPs later. It is simple yet generally endorsed without having any encryption and authentication.

(ii) **Distributed Network Protocol (DNP3)**: More versatile and capable compared to Modbus, DNP can time stamp data, prioritize data, and buffer events. It is widespread in electric utilities and even more so in the secure versions (DNP3-SA).

(iii) **IEC 60870-5-104**: A standard protocol, mainly found in Europe, whose application is in telecontrol. It is suitable in long distance relaying and commonly applied in SCADA systems in the electric grid.

(iv) **OPC-UA**: It is a communication standard independent of the platform, service-oriented, through which secure and interoperable communication is achieved on modern SCADA and IIoT systems.

All the protocols have trade-offs between security and latency and resource requirements. This concurrence in the same SCADA network thereby indicates that the vast majority of CPHS surroundings are hybrid and multilayered.

4.2.4 Data Flow in SCADA and Its Role in Monitoring/Control

The flow of data in the SCADA system is two-directional and cyclical, and includes two steps to acquire, transmit, process, visualize, and execute commands in real time:

Step 1: At the top of SCADA, the array is data capture at the field level, which could include field devices such as telemetry RTU and PLCs data with a sensor detail, such as the temperature, pressure, or voltage. This information can then be sent to an MTU or closed server either by wired or wireless

protocols. On its receipt, the data is processed and analyzed to identify trends, anomalies, and violation of thresholds, visualized in HMIs to raise the operator awareness. Depending on this analysis, whether by automated logic or human action, the result is outputs of the control, commanding it to actuators or other parts of the system, possibly in real time, creating the closed loop that allows efficient and safe operation of systems that is the key to control.
Step 2: This feedback loop is not optional and it is critical to stability of the system, optimization of performance, and real time response. The CPHS version has such a flow also incorporating data logging of transmissions to audit, alerts to human intervention, and AI suggestions of predictive control.

4.2.5 Role of Humans in the Loop: Decision-Making and Interaction

Although automation is fundamentally part of SCADA systems, there is still a need to have human operator involvement—at least in cases or CPHS where safety, ethics, and circumstantial judgment plays its role. People make complicated choices under uncertain circumstances, explain data, and counter-override automated answers where necessary.

Human operators in SCADA systems, in turn, monitor the situation, receive automated commands over HMI and dashboard, activate manual override commands in case of an emergency, and support adaptive control by optimizing AI/ML models using feedback. Digital twins advance their performance even more with the help of training and simulations and make their response even faster and accurate. This human involvement is even more active in CPHS [6]. As an example, it will be possible to operate remotely through mobile apps/augmented reality, and AI can provide explanations to support decision-making. This hybrid interaction is what causes the system to be more resilient but hard to secure and manage.

4.3 Cybersecurity Threat Landscape for SCADA in CPHS

SCADA systems have grown to be highly vulnerable to complex and dynamic cyber related threats as they are presently strongly incorporated in CPHS. These risks are not restricted by external hackers anymore but they vary over a considerable range, involving internal attackers, inadvertent failure, and advanced stealthy attacks. To solve this threat picture, it is important to have proper taxonomy of the kinds of threats and the common attack vectors as well as a study of the practical attacks such as Stuxnet and BlackEnergy. Simultaneously, SCADA systems are also confronted with cybersecurity issues that are special to the field due to older equipment, need for real-time operations, and lack of confidence in separation of the network. In this section, the key areas of threats, procedures that exploit vulnerabilities of SCADA, and challenges systematically experienced in securing these mission-critical systems have been outlined.

4.3.1 Taxonomy of Threats: External, Internal, Intentional, and Accidental

CPHSs with SCADA systems are growing open to a plethora of cybersecurity threats because of their growing amounts of connectivity and their essential importance to infrastructure. The nature of threats may be sorted methodically, according to nature and purpose:

(i) **External threats**: These lie within the outer environment of the organization and in many cases, they are unscrupulous elements such as nation-state hackers, cybercriminals, hacktivists, or script kiddies [7]. These cyber criminals take advantage of vulnerabilities on the network, lack of security of remote entries, or in the supply chain.

(ii) **Internal threats**: Those are carried out by the people working there or are present within the company and are either malicious insiders or negligent employees. An example might be an angry employee introducing malware, or a technician who does not know what he or she is doing misconfigures a device.

(iii) **Intentional threats**: These refer to conscious attacks that are organized to disrupt, steal, or attack. These are usually high-impact measures including the use of ransomware, expert zero-day abuse, or logic rearrangement of industrial systems.

(iv) **Accidental threats**: This type of threat is accidents-driven, through human error, misconfiguration, or even software/hardware errors. Such are common in the metaverse environment by virtue of the complicated mingling of combined legacy apparatus with the current IT infrastructure.

This taxonomy is also necessary because it is crucial in the design of defense in depth frameworks and the integration of intrusion detection and mitigation mechanisms with the nature of the actual incidence.

4.3.2 Common Attack Vectors

A great number of attack vectors exist against SCADA systems in CPHS, and a lot of them are rather well-documented and ever-changing. The most important of them are:

(i) **DoS attacks**: They are directed to interrupt the functioning of SCADA using excessive traffic to overwhelm the operation of servers or communication channels. In industrial settings, DoS may stop the process or cause instability on the grid or endanger the safety systems.

(ii) **MITM attacks**: During such attacks, the adversary intercepts communication between SCADA elements (e.g., between an HMI and a PLC) modifying it (possibly impairing integrity of a communication) and making it seem legitimate to inject malicious commands or manipulated data.

(iii) **Malware and ransomware**: malware such as worms, trojans, or ransomware may enter into SCADA networks through phishing emails or infected universal serial bus (USB) or through compromised updates. HMIs can be corrupted, credentials stolen, or safety systems compromised by these payloads.

(iv) **Protocol abuse**: SCADA protocols used such as Modbus, DNP3, and IEC 60870 are typically not encrypted and authenticated. The attackers have the ability of spoofing commands, reading sensitive data, or replying with legitimate packets, which results in physical damage or loss of data.

(v) **Phishing and social engineering**: These are the cheap and big-profitable attack tactics to human operators, engineers, or just IT personnel. The attackers would be able to gain administrative privileges or to use malware without breaking technical defenses due to manipulation of human behavior.

Such vectors do not exclude each other, and in fact, they are typically phased together in the case of advanced persistent threats in which attackers enjoy persistence over time.

4.3.3 Case Studies of Real-World SCADA Attacks

Several real-world cyber incidents have underscored the vulnerabilities of SCADA systems (Table 4.1) and the cascading impacts they can have on CPHS environments:

These incidents highlight the shift from data-centric threats to attacks on critical infrastructure with real-world consequences.

4.3.4 Challenges in Securing SCADA Systems in CPHS

SCADA systems and protocols are more susceptible to cyber threats due to a few inherent risks, which make IT security solutions hard to apply:

(i) **Legacy systems**: The operating systems (e.g., Windows XP) of a good proportion of SCADA systems are legacy

Table 4.1 Case studies of real-world SCADA attacks and their impact on CPHS Environments.

Attack	Impact	Technique	Relevance
Stuxnet (2010)	Physical destruction of IR-1 centrifuges	Zero-day vulnerabilities, USB propagation, PLC reprogramming	Demonstrated how cyberattacks can lead to physical (kinetic) damage
BlackEnergy (2015)	Mass blackout affecting 230,000 people	Phishing, malware, remote access exploitation	Exposed vulnerabilities in energy sector SCADA and the risks of poor cyber hygiene
Triton/Trisis (2017)	Temporary plant shutdown; potential loss of life	Gained access to engineering workstations	First malware to directly target industrial safety systems

operating systems, which do not natively support current security patches. Such systems were built but not out of cybersecurity.

(ii) **Low bandwidth and computation**: Field devices such as RTUs and PLCs may possess a low computation capacity. This constrains the deployment of a conventional agent-based intrusion detection or encryption algorithm without compromising real-time functions.

(iii) **Real-time constraints**: SCADA works involve deterministic timing. The slightest delay due to any reason, such as security checks or data filtering and encryption, may lead to the instability of the system or its physical failure.

(iv) **Heterogeneity of devices**: SCADA environments encompass multivendor, multiprotocol systems and, thus, standardized security policies are quite hard to put in place [8].

(v) **Air-gap assumptions**: In the past, several organizations made assumptions that the SCADA network was isolated (air-gapped). Due to the emergence of remote access, cloud integration, and IoT, this assumption is no longer applicable, and this leads to an extension of the attack surface on the systems.

The solution to these challenges involves the specific utility of custom security architectures, lightweight IDS/intrusion prevention systems (IPSs), division of the networking form, and human-in-the-loop management to maintain security levels and take better control of cyber resilience.

4.4 Intrusion Detection Systems for SCADA in CPHS

In the face of increased hazards, traditional cybersecurity tools are less able to deal with advanced adaptive threats as SCADA systems are integrated into increasingly more complex CPHS. Traditional measures such as firewalls and signature-based IDS are confined to being quite reactive, and have difficulty in understanding the protocol layer, and do not recognize new or zero-day attacks [9]. This has aroused the need to develop specialized intrusion detection systems specific to the special operation requirement of SCADA within CPHS. Current IDS strategies integrate ML and AI to promptly identify anomalies in the system and pattern recognition, having to face the difficulties associated with the lack of data, their heterogeneity, and the changing behaviors of threats.

4.4.1 Limitations of Traditional Security Measures

Traditional cybersecurity tools and measures, including signature-based intrusion detection systems, firewalls, and antiviruses, have served as the frontiers of SCADA networks. Nevertheless, such heuristics tools have severe limitations in their application to a CPHS-integrated SCADA setting:

The main drawbacks of traditional SCADA security solutions are that they are not adaptable in real time and can easily remain reactive instead of preventing new or zero-day attacks; signature-based IDS make them susceptible to quickly becoming outdated without regular updates, which only allows the detection of known threats; their limited ability to identify the actual anomaly versus their false positive rates causes most basic anomaly detectors to generate too many false positives to process effectively; they are simply unscalable, falling behind the rapidly expanding volume of data and network nodes of modern systems; and they are inclined to show protocol blind spots. To address these shortcomings, there is an indisputable necessity of SCADA-specific, intelligent intrusion detection solutions that would meet the operational and architectural peculiarities of CPHS.

4.4.2 Types of Intrusion Detection Systems

In CPHS environments, intrusion detection systems over SCADA networks may be categorized on the basis of detection strategy, as is shown in Figure 4.1, which spans signature-based IDS that compare known attack

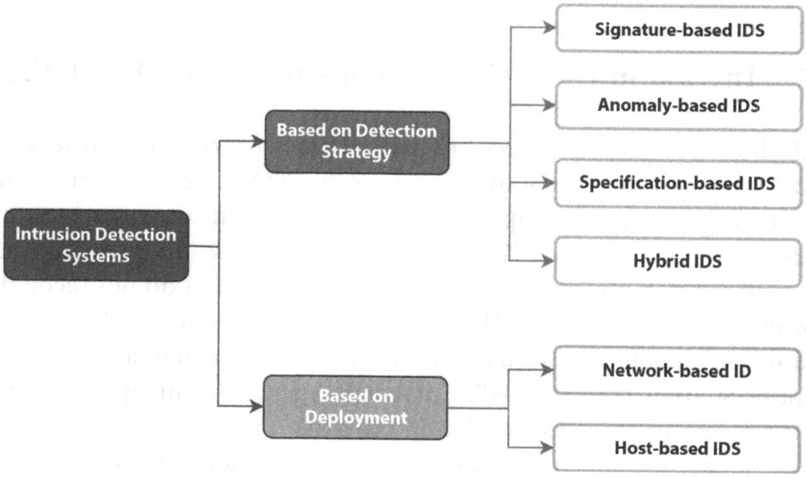

Figure 4.1 Types of intrusion detection.

patterns, anomaly-based IDS that learn normal activity and mark deviations by statistical or AI/ML models [10], specification-based IDS that enforce predefined rules of operation, the hybrid approach that fuses signature and anomaly approaches, and also on the basis of deployment, such as network-based IDS that observe traffic on the SCADA network or host-based IDS that have the potential. In CPHS environments, hybrid and distributed IDS architectures are also becoming more popular as they allow greater visibility across wide areas and anomaly detection on a more local scale in regard to the complex multifaceted relationships between cyber, physical, and human elements.

4.4.3 Role of Machine Learning and AI in Intrusion Detection

AI and ML have transformed the way to design IDS for SCADA due to their ability to design an intelligent system to detect a threat adaptively and less dependent on rules. They mainly play the following important roles:

Pattern recognition: Supervised (e.g., support vector machine, random forest, convolutional neural network [CNN]) ML models can be trained on labeled SCADA traffic to identify the attack patterns.
Anomaly detection: Unsupervised methods (e.g., k-means, isolation forest, autoencoders) can find unidentified threats through the identification of something that is not usual in the mode of functioning.
Time-series forecasting: With some data in SCADA systems being recorded over time (e.g., demand/supply data), recurrent neural networks (RNNs) and long short-term memory (LSTMs) are relevant in modeling such time-series trends to forecast and identify abnormal trends [11].
Federated and transfer learning: These methods enable the decentralization of training and adaptation to various SCADA settings without the centralization of sensitive data.
Online and incremental learning: These are transparently necessary in dynamic CPHS environments, where the behavior of the operation changes and models must respond almost in real time. Due to the latency and compute needs, AI/ML-based IDS can be placed at the enterprise edge or in the cloud. They can also be applied to explainability, which gives human operators an understanding of why an alert was raised.

4.4.4 Dataset Challenges and Training Considerations

ML-based SCADA-specific IDS have a number of challenges related to data; namely, real attack data may not be publicly available and are both confidential and limited, requiring researchers to use means of generating synthetic or simulated data (e.g., Battle of the Attack Detection Algorithms, SWaT, industrial control system), and manual labeling is labor-intensive and requires profound knowledge of the domain. In addition, the attacks presented in an episodic way would lead to class imbalances, which skew the model against benign behavior, and multivariate time-series SCADA data suitable representations have a high number of dimensions, which necessitates feature engineering or dimensionality reduction such as principal component analysis (PCA), t-distributed stochastic neighbor embedding. To make it even more difficult, seeing that different SCADA configurations have different architectures and operations and can be quite dissimilar, models that have been trained to work with one are unlikely to be easily applied to the other [12]. To address such challenges, researchers resort to data augmentation, synthetic attack generation, as well as domain adaptation strategies, and more recently to digital twin simulations to generate realistic, CPHS-targeted training data.

4.5 Advanced Defense Mechanisms in SCADA-Enabled CPHS

The interconnection is growing among SCADA systems in the CPHSs; therefore, defense mechanisms should be more advanced than the conventional security tools. These mechanisms introduce smart analytics, decentralized trust systems, and human controls to generate dynamic and resilient cybersecurity systems.

4.5.1 Behavioral Modeling and Anomaly Detection

Instead of applying rules defined by system, device, and human behaviors, behavioral modeling in the SCADA networks focuses on profiling behavior and making the profile constantly adaptive to several anticipations of behavior. These models identify baselines of normal behavior and outliers through time-series data of sensors, controllers, and HMIs e.g., using PCA and clustering algorithms such as density-based spatial clustering of applications with noise. Anomaly detection can then be applied *via* statistical (thresholds, histograms, distributions), ML (supervised or unsupervised outlier detection), and deep learning models (autoencoders or LSTM networks) to learn complex time structures in multivariate SCADA data. Context-aware detection also enhances this process with environmental, time, and user context information, decreasing false positives, and enhancing interpretation in CPHS settings. When all is said and done, behavior modeling is best suited to the discovery of zero-day attacks as well as stealth incursions that are not so obvious within the normal operations of the organization.

4.5.2 Blockchain for Decentralized Trust and Data Integrity

The blockchain technology hardens SCADA data transactions within CPHS environments by creating decentralized trust, so that all nodes (e.g., sensors or PLCs) can independently confirm commands without relying on a central authority, as well as by providing immutability and tamper-proofing, i.e., pushing event logs and configuration changes into irrevocable and trusted repositories of forensic-quality data [13]. Smart contracts also automate the most important functionality, such as access control and alerting, or machine coordination, as in a lockdown of RTUs when sensor data no longer follow a normal pattern, and lightweight blockchain frameworks that use resources efficiently, such as proof of authority or delegated proof of stake, hold down resource overhead on constrained devices. One example of where blockchain can be used is in smart grids to ensure that the data on energy production and consumption can be tracked securely and ensure that the data are authentic and there is a reduced risk of data manipulation on distributed nodes.

4.5.3 Adaptive and Self-Healing Security Architectures

CPHS-based SCADA systems are built on robust and flexible structures, which incorporate real-time detection and automated mitigation and restoration. IPS augment IDS by dropping offensive packets or isolating problematic hosts; and autonomous reconfiguration provides data rerouting, backup controller switches, or manual-mode failsafe fallback in the face of attack.

Security orchestration platforms are AI-driven: They coordinate the detection, incident response, and policy enforcement at the network layers and on a variety of devices. Continuity is maintained and downtimes are kept to a minimum by self-healing mechanisms, e.g., hardware or functional redundancy, rollback to last known good states, and dynamic patch management. Collectively, these approaches can make a huge difference in mean time to detect and mean time to respond that will enhance cyber resilience in CPHS environments [14].

4.5.4 Human-Centric and Explainable Defense Systems

Human indispensability in decision-making and response is maintained in the CPHS environment, yet human susceptibility to the cognitive mismatch fired back by AI-driven security may increase; explainable security systems will reduce this gap by using explainable AI (XAI) to synthesize interpretable outputs of security solutions based on ML-trained IDS, stating why anomalies or alerts are isolated. The operator-in-the-loop models also help to create the sense of trust calibration, making the alerts neither misheard nor quickly acted upon blindly, and introduce feedback that makes human input improve the IDS models and lessen the false positive rates [15]. Interfaces should ease cognitive burden as they mandate the use of critical alerts, offering insightful information, and making it more actionable; furthermore, human behavior monitoring (e.g., tracking overrides of commands, or logins) can reveal insider threats or stolen credentials. These high-security systems improve transparency, trust, and defense in critical SCADA systems because they sufficiently keep operators engaged.

4.6 Integration with Cloud and Edge Environments

Supervisory control and data acquisition systems grow within CPHSs and beyond, making the combination with cloud and edge computing and fundamentally changing the ways the data are processed, stored, and secured. Such paradigms introduce scalability and flexibility to applications, as well as new architectural and security considerations.

4.6.1 Cloud Integration: Benefits and Security Implications

Integration by the cloud enables the SCADA-enabled CPHS with the ability to aggregate and centralize the data, scale to store the massive time-series streams received by RTUs, PLCs, and sensors, and allows performing the advanced analytics of ML/DL models all in the cloud, as well as secure global remote monitoring and deployment of patches and updates [16]. But this development also increases the attack surface of systems, suggesting vulnerability to phishing, credential theft, and brute-force attacks, and necessitates strong cryptography and integrity verification of the data in motion and at rest. Using third-party cloud providers poses supply-chain and shared-responsibility issues and processes performed in real-time need to be compensated by delays in cloud processing. In a desire to maintain a balance between capability and security, a number of deployments have taken the hybrid approach to maintaining the sensitive control functions on-premise and offloading the analytics to the cloud.

4.6.2 Edge Computing for Real-Time Intrusion Detection

The edge computation brings computation nearer to the data source, i.e., PLCs or field gateways, and thus it is quite accommodating to latency-sensitive applications in CPHS.

Benefit to SCADA: The work of local processing of intrusion detection and control logic using edge computing improves SCADA-enabled CPHS by increasing the responsiveness in time due to lower latency. It gives the capability to do filtering and preprocessing on the source, not sending out unnecessary data, thus still saving on bandwidth. Through edge-based IDS/IPS, anomaly responses would be immediate with no dependency on central systems [17]. In addition, edge nodes are independent when an Internet connection fails or during a server attack and this enables them to continue functioning under unfavorable circumstances.

Edge security: The SCADA-enabled CPHS frequently use edge devices to host lightweight AI models e.g., a decision tree or a small CNN on an edge device to do real time analytics locally, and include trusted execution environments to ensure that such computations cannot be read by unauthorized users. Blockchain- or hash chain–based technologies verify data generated at the edge and regulate logs en route to distribution upstream, integrating low-overhead intelligence with high-assurance security, enabling robust, secure, and trustworthy edge operations.

Use case: The concept of an attack technique such as a command injection attack may be detected in real time using edge analytics on a robotic assembly line in smart manufacturing and cascaded with a subsequent automatic shutdown without requiring cloud connectivity.

4.6.3 Hybrid Architectures for SCADA Security

To improve their performance, resilience, and security, a hybrid SCADA architecture can use both edge and cloud resources by applying distributed IDS/IPS frameworks, with detection of anomalies at the edge supported by threat intelligence sharing done in the cloud, and exploiting tiered data processing, where the most time-sensitive measurements are processed at the edge, and the historical data analysis along with predictive modeling runs in the cloud. Unnecessary control paths allow an alternative toward manual or semiautonomic controls in case cloud services are compromised, and consistency of security policies is ensured as unified security policies are planned at a central point and simultaneously imposed across all the edge nodes [18]. Privacy-sensitive federated learning (FL) allows training AI intrusion–detection models collaboratively across the devices without sharing raw data, and all communication between edge, cloud, and SCADA components is secured using secure application programming interfaces. The combination of these qualities into a hybrid solution can fulfill high demands of CPHS environments in terms of real-time, scalable, and resilient environments.

4.6.4 Challenges in Multi-Cloud and Multi-Edge Security Coordination

Multi-cloud and multi-edge adoption bring their own problems to the process of maintaining a coherent security stance: they can provide uneven visibility across varied terrain, potentially causing network activity to go unnoticed, introduce gaps and inconsistencies about incident response, and require highly specialized tools to authorize, encrypt, and control access across a host of disparate operating systems [19]. Also, the data sovereignty and regulatory compliance (e.g., General Data Protection Regulation, Health Insurance Portability and Accountability Act) of geographically distributed clouds and edge nodes are hard to implement, and threats may quickly spread across the connected zones unless tightly segmented. The approaches of countering these problems include the deployment of centralized cross-domain SOAR (security orchestration, automation, and response) platforms, the introduction of zero trust architectures to remove periods of implicit trust, and secure containers and virtualization to isolate workloads across several platforms.

4.7 Human Factors and Usability in Intrusion Detection

Although most of the SCADA security in CPHS considers the technical aspect of the security, the human factor is the most important aspect of the security that is ignored. Engineers, administrators, and human operators are highly involved in the SCADA processes. As a result, there is more of a need to not only come up with intrusion detection systems that are precise, but also usable, interpretable, and conform to the patterns of human decision-making to be relevant to security.

4.7.1 The Role of Human Operators in SCADA Security

In SCADA-equipped CPHS, human operators primarily are charged with the responsibility of following up all alerts, confirming the healthiness of the system, and reacting to the breaches, but it is absolutely essential that they remain within a manageable cognitive load that a combination of unnecessary or too-involved alerts may leave them unable to respond except by letting them go and that they have such confidence in the IDS as makes obscure or repetitive alerts seem safe to ignore. Their reaction time is crucial, because when it comes to misperceiving an alert or failing to comprehend it and take actions immediately, the mitigation opportunities are lost, and their response has to correspond to the secured procedural routine and standard operating procedures integrated into industrial processes. Because cyberphysical anomalies in high-consequence CPHS environments have physical impacts, human-aware, interpretable security systems are thus a necessity to maintain operators and make deliberate decisions with confidence and in real time.

4.7.2 Human-in-the-Loop and Human-on-the-Loop Models

Intrusion detection in SCADA CPHS has increasingly adopted human-centered control architectures: human-in-the-loop models, which prioritize transparency and the provision of context-dense alerts and whose IDS input must be actively interpreted by operators and decisions made; human-on-the-loop (HOTL) systems, which operate independently with human oversight and where human intervention is only provided in response to anomalies; and entirely autonomous human-out-of-the-loop systems, which deal with time. HITL and HOTL are the best compromise in SCADA, where there is still some activity by the operator, and it reaches the best situational awareness and accountability, but relies on automation speed and efficiency.

4.7.3 Designing Usable Intrusion Detection Systems

The capabilities of a technically advanced intrusion detection system are of hardly any importance when such a system cannot be comprehended or even trusted by its operators, and therefore usability is an important aspect to consider during the process of developing the design. The design of effective IDS must focus on alert categorization and use their severity to create groups to alleviate alert fatigue and provide better reactivity. A dashboard that gives users an intuitive view of the system state, is presented visually, e.g., through network graphs and time-series plots, to help convey alerts and system state effectively. The inclusion of natural language explanations by using XAI will improve the level of user comprehension of the reasons behind certain alert triggers. Such interactive feedback using false positives or authentic threats also enables ongoing ML model learning [20]. Moreover, the personalization of the interfaces to various user groups so that engineers demand the details that the technical team could provide as well as the chief information security officers who need the high-level overview of the systems makes sure that the system fits a wide range of operations. After all, when it comes to human perception, cognition, and decision-making, it is perfectly reasonable to tune in the IDS capability to that of a human being, allowing the user to act appropriately and effectively during such pressure.

4.7.4 Training, Awareness, and Organizational Readiness

The security of SCADA systems within the CPHSs cannot be provided by means of technical tools only; the preparedness of people plays a crucial role as well. Extensive security training is critical to ensure that every member of the personnel is familiar with the cyber risks, operation of SCADA, and how to react to the alerts or other anomalies satisfactorily. Cyberphysical drills and national security exercises can be run regularly to help teams practice and hone their response plans and test using digital twins or testbeds. A deployment and control of IDS solutions will also need cooperation between cross-disciplinary teams and groups such as cybersecurity professionals, SCADA engineers, and process specialists. Furthermore, IDS-derived alerts need to be incorporated into the larger organizational policies such as incident response systems, compliance regimes, and audits. The establishment of a healthy security culture across the organization means that all people including technicians and executives are on the same side, moving together toward the objectives of comprehensive defense of the physical and cyber space.

4.7.5 Ethical and Privacy Considerations in Human Monitoring

Intrusion detection systems have the possibility of monitoring even human behaviors such as human commands and access practices in CPHSs. This is important in identifying the insider threats; however, this comes at the cost of serious ethical and privacy implications. It should be transparent—operators should know what and why they are being monitored. This data gathering should also be within the principle of minimization of data i.e., only data relevant to behavior shall be collected. Specifically, in such sensitive areas as healthcare and critical infrastructure, where the personal data concern is involved, consent and adherence to the legal and regulatory regimes are particularly imperative. Moreover, the IDS based on AI has to be data-intelligent to eliminate any form of bias to eliminate discrimination and unfair profiling. Because the question of surveillance has increasingly become complex in CPHS, human dignity, trust, and accountability are required to make technological progress ethical in relation to human values.

4.8 Case Studies and Real-World Applications

This section contains real-life scenarios in which intrusion detection systems have been used or could have helped to reduce threats in SCADA-enabled CPHS. These case studies emphasize the real-life difficulties and possibilities of developed cybersecurity mechanisms in the real-world scenario.

4.8.1 Industrial Control Systems—Stuxnet and Beyond

The next example of case studies demonstrates the changing environment of intrusion detection systems in cyberphysical and SCADA settings in different industries. The 2010 Stuxnet malware has been treated as a milestone of the cyberattack on the Iranian nuclear plant, using the vulnerabilities of Siemens PLCs, zero-day flaws, code injection, and fake HMI interfaces. It did physical harm but did not induce alarms, which shows the significance of behavioral anomaly detection and how one can use the trust in operator interfaces. The attack on the power grid of Ukraine in 2015 provided evidence of the disastrous capabilities of cyberattacks in smart grids. Spear-phishing and RATs (remote access trojans) gained the hackers access, as the hackers remotely opened circuit breakers and disabled recovery systems, disrupting services of 230,000 residents [21]. It was a case that puts multilayered IDS, HOTL supervision, and integrated cyberphysical response on emphasis. Table 4.2 illustrates key risks and IDS solution strategy for the lessons learned across different domain sectors.

In a research use case, the SWaT testbed demonstrates a practical test environment to test IDS in water treatment systems with real PLCs, labeled data, and attacking capabilities, such as sensor spoofing, and process tampering, closing the gap between laboratory experiments and deployment

Table 4.2 Lessons learned across domains.

Sector	Key risk	IDS solution strategy
Power grids	Remote command manipulation	AI-based anomaly detection and response automation
Manufacturing	Insider and zero-day threats	Behavioral modeling and smart contracts
Healthcare	Device spoofing and data leaks	Edge IDS with secure firmware analytics
Water systems	Valve/sensor tampering	Digital twin modeling for preattack simulation

on real plants. In the sphere of healthcare CPHS, the remote monitoring devices and follow-up skills, with older firmware, inadequate verification, and low-quality encryption, are profound safety problems. IDS in such settings aid behavior profiling, abnormality alerting, and edge-based detecting protocols such as HL7 (Health Level 7) and DICOM (Digital Imaging and Communications in Medicine) to safeguard patient safety. Finally, the security issues of industrial automation were displayed in the 2018 opportunity of a U.S. natural gas pipeline operator. There was a breakdown of communications between the controllers and field devices as a result of attackers and this was made worse by improper segmentation. The use of protocol-aware IDS and edge analytics, as well as novel solutions, such as the use of blockchain to validate control logs, have now become the major trends in the security of such infrastructure. These cases combined highlight the problem of the multidimensional purpose of IDS in the security of CPSs in industries.

The practical cases confirm that intrusion detection in SCADA-based CPHS has to be adaptable, domain-specific, and human-sensitive. In addition, ongoing feedback on incidents in operation is essential in the design and deployment of the IDS strategy.

4.9 Future Directions and Research Challenges in SCADA Security for CPHS

As the SCADA networks are still expanding within the CPHSs, there is an even more pressing need for advanced and evolving intrusion detection systems. The emergence of interdependent environments has established new threats such as AI-based cyberattacks, supply chain risks, and cross-domain attacks, all of which can only be prevented by IDS that is able to detect intelligent and coordinated attacks.

In addition to these improvements, ML/AI-based IDS still face fundamental limitations—a lack of a large number of real-world datasets, scalability, high false-positive probability, and a lack of model transferability. IDS of the future should be self-tuning, environmentally sensitive, as well as able to adjust to various operational environments [22]. FL has a possible future because it allows privacy-preserving collaborative training of models on distributed edge devices, although it will have to address such problems as data heterogeneity and data poisoning attacks [23].

XAI can play a crucial role in establishing the trust of complex models of an IDS, as operators can figure out alerts to meet the deadlines mentioned in audits. In the meantime, a quantum computer poses a threat to current cryptographic standards, which requires the use of postquantum and lightweight encryption in the SCADA systems. Growing in popularity are digital twins, a tool that has simulations that model an attack and improve anomaly-catching security predictions, but the ability to be effective at real-time predictive security is still a problem [24].

This requires human-centered design, and IDS interfaces designed in such a way that they conform to the roles of the user and the cognitive load they have to perform [25]. A lack of regulatory consistency across business communities and territories is another regulator of the improvement of cybersecurity, and there is a need to establish uniform principles and international coordination. In future visions, there is a wish for self-healing, intelligent, autonomously operated security systems that use combinations of AI, blockchain, FL, and digital twins to protect SCADA systems in unpredictable and hostile environments.

4.10 Conclusion and Summary

The modern interconnected world has made SCADA networks the backbone of CPHSs, enabling critical infrastructure in different industries to operate inside domains such as energy, healthcare, and transportation. Because such systems become more integrated and multifunctional with technologies such as IoT, AI, and cloud computing, they have become much more vulnerable to cyberattacks. This chapter presented the discussion of the use of intrusion detection systems in SCADA networks security with references to architectural implications, threat landscape analysis, and complex detection tools, such as the use of ML, blockchain, FL, and digital twins.

Human-centered design, explainability, and decision support were highlighted as the crucial components of the contemporary IDS, particularly in a high-stakes setting where the operators continue to act as crucial decision-makers. Examples of case studies such as Stuxnet and the attack on the power grid of Ukraine reflected the consequences of an attack in real life and the necessity of actions to prevent it. In the future, it is crucial to come up with self-adaptive, explainable, and cryptographically robust IDS solutions. This global standardization and regulatory consistency are also important. Finally, protecting SCADA-based CPHS is not merely a tech-related issue but a crucial issue facing society, as a matter of cooperation between disciplines and sectors.

AI disclosure statement: The author confirms that no artificial intelligence (AI)-generated content or AI-assisted writing tools were used in the preparation of this chapter (including summaries and the abstract). All text, analysis, and conclusions are the result of the author's original work.

References

1. Lee, E.A., The Past, Present and Future of Cyber-Physical Systems: A Focus on Models. *Sensors*, 15, 4837, 2015.
2. Rajkumar, R., Lee, I., Sha, L., Stankovic, J., Cyber-Physical Systems: The Next Computing Revolution. *In Proc. Des. Autom. Conf.*, 47, 731, 2010.
3. Galloway, B. and Hancke, G.P., Introduction to Industrial Control Networks. *IEEE Commun. Surv. Tutorials*, 15, 860, 2013.
4. Mohan, N., Ravi, V., Manogaran, G., Role of SCADA in Cyber-Physical Systems: A Comprehensive Review. *J. Ind. Inf. Integr.*, 27, 100286, 2022.
5. National Institute of Standards and Technology (NIST), Guide to Industrial Control Systems (ICS) Security (NIST SP 800–82 Rev. 2), 2015.
6. Krook, M. and Baskar, S., Role of Humans in Cyber-Physical Production Systems: A Survey. *Comput. Ind. Eng.*, 161, 107601, 2021.
7. Alcaraz, C. and Lopez, J., A Security Analysis for Wireless Sensor Mesh Networks in Highly Critical Systems. *IEEE Trans. Syst. Man Cybern. Syst.*, 40, 419, 2013.
8. Cárdenas, A.A., Amin, S., Sastry, S., Research Challenges for the Security of Control Systems. *Proc. USENIX HotSec.*, 3, 1, 2008.
9. Ghafir, I., Prenosil, V., Hammoudeh, M., Jabbar, S., Baker, T., Jaf, S., BotDet: A System for Real Time Botnet Command and Control Traffic Detection. *IEEE Access*, 6, 38947, 2018.
10. Giraldo, J., Sarkar, E., Cárdenas, A., Maniatakos, M., Kantarcioglu, M., Security and Privacy in Cyber–Physical Systems: A Survey of Surveys. *IEEE Des. Test.*, 34, 7, 2018.

11. Knowles, W., Prince, D., Hutchison, D., Jones, K., Janicke, H., A Survey of Cyber Security Management in Industrial Control Systems. *Int. J. Crit. Infrastruct. Prot.*, 9, 52, 2015.
12. Kumar, P., Vealey, T., Srivastava, H., Security in Internet of Things: Challenges, Solutions and Future Directions. *Proc. Hawaii Int. Conf. Syst. Sci.*, 49, 5772, 2016.
13. Lee, R.M., Assante, M.J., Conway, T., Analysis of the Cyber Attack on the Ukrainian Power Grid. *SANS Ind. Control Syst.*, 1, 38, 2016.
14. Miller, B. and Rowe, D., A Survey of SCADA and Critical Infrastructure Incidents. *Proc. Annu. Conf. Res. Inf. Technol.*, 1, 51, 2012.
15. Mitchell, R. and Chen, I.R., A Survey of Intrusion Detection Techniques for Cyber-Physical Systems. *ACM Comput. Surv.*, 46, 29, 2014.
16. Kaur, P.D. and Chana, I., Cloud based intelligent system for delivering health care as a service. *Comput. Methods Programs Biomed.*, 113, 346, 2024.
17. Enam, M.R., Joarder, M.M.I., Taimun, M.T.Y., Sharan, S.M.M.I., Framework for Smart SCADA Systems: Integrating Cloud Computing, IIoT, and Cybersecurity for Enhanced Industrial Automation. *Saudi J. Eng. Technol.*, 10, 5, 2025.
18. Slay, J. and Miller, M., Lessons Learned from the Maroochy Water Breach. *Crit. Infrastruct. Prot.*, 253, 264, 2007.
19. Teixeira, A., Shames, I., Sandberg, H., Johansson, K.H., A Secure Control Framework for Resource-Limited Adversaries. *Automatica*, 51, 135, 2015.
20. Liu, Y., Yu, W., Ai, Z., Xu, G., Zhao, L., Tian, Z., A Blockchain-Empowered Federated Learning in Healthcare-Based Cyber Physical Systems. *IEEE Trans. Netw. Sci. Eng.*, 10, 2685, 2023.
21. Alam, K., Kifayat, K., Sampedro, G.A., Karovič, V., Naeem, T., SXAD: Shapely Explainable AI-Based Anomaly Detection Using Log Data. *IEEE Access*, 12, 2024.
22. Holzinger, A., Carrington, A., Müller, H., Measuring the Quality of Explanations: The System Causability Scale (SCS). *KI - Künstl. Intell.*, 34, 193, 2020.
23. Li, T., Sahu, A.K., Talwalkar, A., Smith, V., Federated Learning: Challenges, Methods, and Future Directions. *IEEE Signal Process. Mag.*, 37, 50, 2020.
24. National Institute of Standards and Technology, Post-Quantum Cryptography Project, 2023.
25. Mliki, H., Kaceam, A., Chaari, L., A Comprehensive Survey on Intrusion Detection based Machine Learning for IoT Networks. *ICST Trans. Secur. Saf.*, 8, 171246, 2021.

5

CPHS–IoT Models: A Hybrid Approach to Deal with Engineering Systems and Control Modules

L.K. Indumathi[1]*, Merugu Praveen Kumar[1], Sonal Bhide[1] and P. Manjula[2]

[1]Department of Computer Science and Engineering, Matrusri Engineering College, Hyderabad, Telangana, India

[2]Department of Health Information Management and Technology, College of Applied Medical Sciences, King Faizal University, Alahsa, Ash Sharqiyah, Saudi Arabia

Abstract

Globally, significant outlays have recently been made in the growth of cyberphysical human systems (CPHSs) as alternatives to important economic hardship challenges. The capacity of the Internet of Things (IoT) to enhance current CPHS platforms with recognizing abilities has also led to its broad adoption. Nonetheless, consideration needs to be given to how connected devices affect the reliability of CPHS platforms. Using an interference-aware network redesign and a mass model of the underlying dynamics within the CPHS–IoT, these data tackle the reliability problems with CPHS. Sensors enhance the safety of CPHS architecture by preventing traffic instability through effective monitoring. An innovative method for developing systems and control modules is the combination of the IoT and CPHS paradigms. To create smart, adaptable, and real-time control systems, this chapter explains a fusion modeling basis that syndicates CPHS with IoT technologies (CPHS–IoT). This chapter presents a hybrid modeling framework that combines CPHS with IoT technologies. Also, this chapter offers research problems, implementation case studies, architectures, hybrid modeling approaches, and fundamental ideas. For scholars, researchers, and system engineers investigating applications of intelligent engineering, the chapter seeks to offer a comprehensive overview.

***Keywords*:** Cyberphysical human systems, control, Internet of Things, applications, networks structure, systems, cyberphysical human systems–Internet of Things

**Corresponding author*: indumathibiet@gmail.com; ORCID: 0000-0001-9873-9791

Anitha Velu, Prasanth Aruchamy, Raghu Ramamoorthy, Rajesh Kumar Dhanaraj and Manjula Arunraj (eds.) Cyber Physical Human Systems: Emerging Trends and Applications, (97–130) © 2026 Scrivener Publishing LLC

5.1 Introduction

The world is changing far more quickly than it has in the past. New perspectives on science, technology, and social trends are desperately needed now more than ever before. Both the population and the use of the planet's resources are rapidly increasing. According to estimates, there are over eight billion people on the planet, and each person uses around 20 megawatt hours of energy per year, or almost 2 tons of oil equivalent. Thankfully, as strong communication infrastructure has grown and become more accessible, so too has an understanding of the world's situation and the capacity to affect it.

5.1.1 Cyberphysical Systems and Hybrid Systems

Although the literature uses a variety of definitions, the following early definition is sufficiently straightforward and obvious. The National Science Foundation in the United States [1] first used the term "cyberphysical systems" (CPSs) in [2] to describe the fusion of physical and computational processes and components as well. Certain technical needs become apparent when examining this description from the perspective of the mathematical concepts required to model them: Discrete modeling is required for computational components, continuous modeling is required for physical components, and probabilistic and perhaps game theoretic modeling is required for telecommunications and networks elements.

In several fields of computational modeling, either continuous or discrete mathematics will be used instead of being combined. In actuality, combining these two may result in some serious technological issues. An extra degree of caution is needed when adding probability on top. On a more pragmatic level, however, a modeling formalism that can express both discrete and continuous components is necessary only to define models that include each [3]. This is exactly what continuous/discrete systems that are hybrid offer. To represent CPSs, these data will heavily utilize continuous, discrete, and blended systems in this work. Also, the authors should utilize probability to introduce fundamental ideas in game theory, interaction monitoring, and movement [4].

Conventional control methods can no longer keep up with the level of complexity of today's engineering frameworks. These systems require human participation in control loops, real-time responsiveness, flexibility in response to changing circumstances, and smooth component-to-component communication. To meet these demands, the Internet of Things (IoT) and cyberphysical human system (CPHS) provide complementary advantages [5]. To accomplish multiplicative relations, CPHS combines human, physical, and computer components. By establishing a means of communication link between gadgets, sensors, and systems as a whole, the IoT makes a contribution. The hybrid CPHS–IoT paradigm makes use of both the IoT's connectivity and sensing capabilities as well as CPHS's compute and control features. This hybrid approach is especially useful in fields such as adaptive buildings, medical care, autonomous systems, and smart manufacturing.

5.2 Basics of CPHS and IoT

Even though the phrase CPSs is relatively new, it is mandatory to think about whether the field or concept underlying this method of system analysis is as well. It is critical that leaders, inventors, and innovators in a subject consider and comprehend the nature of the field and its motivations [6]. One must take into account a variety of related disciplines and think about their connections when one accomplishes this for CPS. The given section of this chapter examines a number of pertinent ideas and talks about how they relate to CPS.

A precise model that displays instantaneously nonstop and periodic features is termed a fusion system; allied fields contain swift comparisons of signal and converting erections. It is crucial to differentiate between the idea of the real, tangible system and its numerical representations, as well as the methods used to investigate them, even though many CPSs can be theoretically represented as hybrid systems. The purpose of mathematical models is to describe observed behavior; yet, they must be updated or even altered when new data arise that the model is unable to explain. It is a common misunderstanding to consider computational system models to be discrete and physical system models to be continuous. Physical system models might be discrete, continuous, or a mix of the two.

For instance, the foundation for quantum mechanical models was laid by Bohr's 1913 model, which proposed that electrons could only exist in distinct orbits around an atom's nucleus. Also, this hybrid model is like a bouncing ball, where bouncing is a periodic event and flying is a continuous behavior. Moreover, many significant events at the level of the quantum are stochastic systems rather than only continuous or discrete systems [7].

It is not consistently feasible to think about systems for computation as strictly discrete systems. Generally speaking, digital machines are built as continuous electronic circuits that are only reliable when considered as separate components. Additionally, there are continuous systems referred to as analog computers. Digital machines can be used to precisely realize several examples of such systems. Lastly, a current field of study that uses probabilistic models is quantum computing. Whereas hybrid systems are a mathematical concept, CPSs are actual objects.

A technology that is integrated into a physical system is called an embedded system. There is an embedded system in every CPS. The vital variance is that the phrase embedded system emphasizes the computing component, which is a part of a broader, physical system. Development on systems that are embedded has historically concentrated on issues including equipment design, power and manufacturing cost reduction, embedded software advancement, and formal verification of discrete systems (automata) [8]. The CPS perspective highlights in what way critical it is to deliberate on the computational system's concern situation, which is frequently required to design, test, and validate the capabilities that are created.

A system that operates in real time is one that has to react to changes in the outside world within predetermined time limits. Some real-time systems have embedded components, but several are not. For instance, even if a computerized trading agent needs to work under tight time limitations to respond effectively to quickly shifting market conditions, it would not often be considered an embedded system. Sequencing or real-time tasks in systems with regular or irregular demand outlines, many computational possessions, tasks of diverse significances, and ongoing interaction have historically been the emphasis of concern systems work. It goes without saying that weak scenario time considerations are often the focus of this field's study. A machine that operates in real time may or may not be a CPS: Real-time limits are present in the car's control system, although they are not always present in the auditory machine.

The dimensions of a structure to function even when some of its mechanisms fail are represented as reliability. Manufacturing parts from stronger materials is the first step toward achieving dependability [9]. Other strategies include adding duplication and testing to identify and effort to adjust for mistakes and failures. Probabilistic techniques have been successfully used in numerous fields, including systems for computation, to improve reliability while controlling costs. But statistical approaches are but one instrument for creating dependable systems. Gaining proficiency in other more basic systems design ideas enables us to apply probabilistic techniques more skillfully. Every CPS has different requirements for dependability, and not every CPS's development must prioritize reliability [10]. Nevertheless, generally speaking, the necessity to take into account the consequences for system-wide integrity for every kind of distinct cyberphysical component grows along with the connectivity between various CPSs.

A more comprehensive concept, dependability can include a number of related qualities, including availability, durability, safety, security, integrity, and maintainability [11]. It is seen as a device of these collective potentials in the context of the multidisciplinary zone of systems manufacturing. Because of its distinct perspective between traditional engineering and traditional management, systems engineering is often viewed as a branch of both engineering and engineering management. It was supposed that reliability and concern engineering are crucial for trendsetters and inventors, and that the given information will offer the reader a sturdy base for further research in these areas.

A mathematical model with interactive elements is defined as a multiagent system. It is regularly linked to game theory, a branch of mathematics. Game theory and multiagent systems offer helpful methods for simulating and reasoning about ideas such as cooperation, competitiveness, belief, knowledge, and intent. There are mathematical models of discrete and hybrid (continuous/discrete) games, and concepts of intelligence are frequently examined in connection with multiagent schemes. Game theory and multiagent systems research essence is on the behavior of collections of agents somewhat more than discrete actors, in contrast to the study of hybrid systems. Lastly, the fields of robotics, control theory, mechatronics, and the IoT are all closely associated with CPS [12].

These fields all have the trait of being extremely multidisciplinary. CPS is comparable to the last discipline (e.g., IoT) and can be seen as an attempt to adopt an even more comprehensive approach than the previous three. The foundations of hybrid systems and communications and networking may always support smart agriculture [13].

5.2.1 Cyberphysical-Human Systems

Human aspects are explicitly incorporated into system modeling and control in CPHS, an extension of classical CPS. In CPHS, people engage with systems as both users and components that both affect and are affected by system states. One of the characteristics of CPHS is the incorporation of human cognitive models into control algorithms. Feedback loops that take human input and system dynamics into account. These data find applications in modern robotics, telemedicine, and autonomous transportation [14].

5.2.2 Internet of Things

The IoT is the technological foundation of contemporary layered networked systems [15]. It permits distributed monitoring and control by allowing physical devices to exchange data *via* the Internet. IoT elements:

- **Sensing layer**: Consists of radiofrequency identification (RFID), sensors, and physical parameter monitoring devices.
- **Network layer**: Uses protocols such as message queuing telemetry transport (MQTT), hypertext transfer protocol (HTTP), and constrained application protocol (CoAP) to control consistent data transfer [16].
- The data layer practices edge or cloud computing to supply and development statistics.
- The application layer enables handler contact and makes analytics, notices, and mechanization possible [17].

5.2.3 Inspiration for Hybrid CPHS–IoT Models

The requirement to progress smart, flexible, and human-centered systems is what stimulated the expansion of hybrid CPHS–IoT models. Real-time decision-making is made possible by these models, which combine human interaction, computational intelligence, and physical instruments. They improve system efficiency, safety, and personalization by fusing IoT connectivity with human and cyber components. The increasing need for automation that is nevertheless sensitive to human values and input is what motivates these hybrid approaches. Their ultimate goal is for people, machines, and the digital world to work together seamlessly. When CPHS and IoT are combined, certain restrictions are addressed separately: comprehensive real-time data connectivity is absent from CPHS alone. IoT is devoid of human-in-the-loop adaptability and sophisticated control systems. When combined, they make context-aware, customized, and self-governing control systems possible [18].

5.3 Hybrid CPHS–IoT Modeling Framework

A hybrid modeling structure for the IoT and CPHS provides a thorough method for handling the complexity of networked resources, including both human actors and IoT devices [19]. To deliver even movement of data and facility planning, this construction makes a habit of source conceptualization, layered strategy, hybrid modeling methods, application-independent arrangement, and message-driven connections. Participating anthropological factors, recognizing applicants as data creators and clients, and joining to matters such as ease, discretion, and safety are all crucial zones of attention. To guard delicate statistics and confirm compliance with guidelines, it also provides a big priority on robust safety and confidentiality actions, such as cryptography and authorization. Institutions or organizations can advance suppleness, scalability, competence, and cost-effectiveness throughout the range of applications, with smart cities, healthcare, and industrial automation, by using such a framework [20].

A hybrid CPHS–IoT architecture comprises various interrelating layers:

- **Application layer**: Provides functions such as visualization, artificial intelligence (AI)–powered decision-making, and cloud dashboards.
- **Cognitive-human layer**: This layer simulates people's responses, choices, and intellectual responses.
- **Cyber layer**: Performs computations, evaluates information, and makes machine learning (ML) inferences.
- **Physical layer**: This layer is composed of mechanical/electrical systems, controls, and devices.
- **Communication layer**: The communication layer makes sure all information is transferred among sensors, devices, and systems.
- **Sensor layer**: Gathers system and environmental information.

5.3.1 Application Layer

In a hybrid CPHS–IoT design, the application layer is crucial for combining everyday items and enabling connection with information technology, user data, and backing up tangible systems. This layer, which is at the highest point of the layout, acts as a link between the people who utilize the structure and its intellect. The application layer serves as a bridge between consumers and the intelligent system underlying a hybrid CPHS–IoT design. It offers processed information from the cyber, human, and physical components in an understandable and practical style. This layer provides immediate overviews, alerts, and control options to assist clients in efficiently monitoring and managing connected gadgets. By combining data analytics and consumer input, it also makes decision-making easier. The application layer provides context-sensitive and customized services that enable smooth human–system engagement. All things taken into account in a CPHS–IoT context, it is the main enabler of user-centered functionality. The procedures and their descriptions are represented in Table 5.1.

The schematic shows the layered architecture of a network of things-based system that integrates human activity and data analytics. The physical IoT layer is made up of sensors and other gadgets that gather data

Table 5.1 Functions of application layer.

Function	Description
Interface for end-users	Affords user screens, mobile apps, website stands, etc., for observing and regulatory CPHS apparatuses
Custom service conveyance	Customizes outputs (decisions, alarms, and visualizations) for certain industries, such as intelligent production, urban planning, or healthcare
Interrelates with Human ideas	Uses UI/UX techniques to record customer preferences, suggestions, and manually overrides
Obtains managed data	Obtains aggregated or real-time visions from the data layer to ease activities or decision-making
Achieves application sense	Encompasses commercial rubrics, choice reason, and artificial intelligence (AI)–driven sanctions
Safety and confidentiality execution	Uses authentication procedures, information confidentiality guidelines, and user-level access regulator
Interoperability layer	As required, integrates with third-party application programming interface (APIs), Supervisory Control and Data Acquisition (SCADA), and Enterprise Resource Planning (ERP) systems

related to the environment. These data are transmitted *via* the networking and communication layer, which ensures consistent connectivity among system and device components. The cyber–human and data analytics layer processes the information, performs analytics, and facilitates the involvement of humans in decision-making. These insights are used by the application layer to provide end users with intelligent services and applications. Lastly, users may communicate with the system, view real-time data, and get alerts thanks to the GUI/notification interface. These layers work together to form an all-encompassing and effective IoT architecture that facilitates intelligent, data-driven operations. Figure 5.1 explains the communication between various layers based on the application layer.

Through sophisticated interfaces that allow for user interaction, monitoring, and control, the application layer is essential in a variety of disciplines. In the medical field, it facilitates telemonitoring dashboards that notify caregivers of any noteworthy alterations in a patient's vital signs.

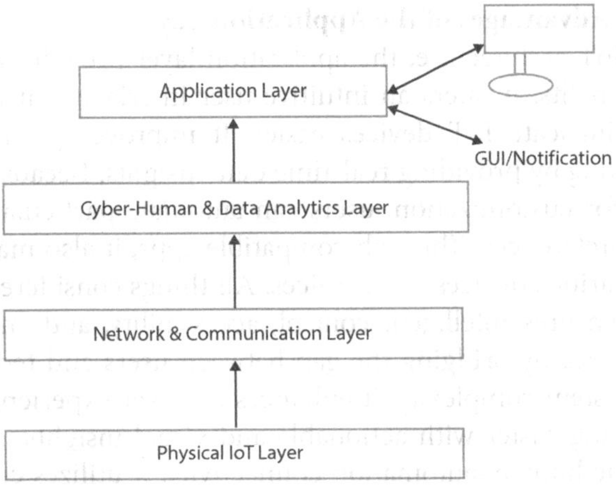

Figure 5.1 Data flow in application layer.

Gateways that enable users to track energy usage and establish preferences to encourage energy saving are offered by the smart grid industry. To guarantee operational efficiency, Industry 4.0's application layer provides interfaces that show the health condition of devices and notify users of planned maintenance. IoT sensor data and user preferences are used by mobile applications in smart homes to automatically regulate air conditioning, lighting, and security systems. The application layer improves automation and decision-making across several domains by bridging the gap between analysis of data and end-user interaction. Table 5.2 explains various application layer examples based on various domains.

Table 5.2 Various domain of application layer.

Domain	**Application layer example**
Healthcare	Dashboards for telemonitoring that notify carers of changes in a patient's vitals
Smart grid	Gateway for users to track energy usage and establish preferences for energy conservation
Industry 4.0	Interface that provides scheduled service notifications and displays the health of the device
Smart home	A mobile application that uses user preferences and Internet of Things sensors to operate security systems, lights, or air conditioning

(i) Principal Advantages of the Application Layer

In a CPHS–IoT architecture, the application layer provides a number of significant benefits. It offers an intuitive user interface that makes interacting with intricate IoT devices easier. It improves productivity and decision-making by providing real-time data insights. Because of this layer's support for customization, users can automate particular operations and specify preferences. Through compatible apps, it also makes it easier to integrate various devices and services. All things considered, the application layer ensures intelligent control, accessibility, and enhanced system performance by bridging the gap between users and technology. By abstracting system complexity, it enhances the user experience. It makes decision-making easier with actionable and visual insights. Also, it provides real-time human–automation connectivity. It utilizes custom application services to generate value specific to a given domain.

(ii) Cognitive-Human Layer in a Hybrid CPHS–IoT Architecture

To direct system behavior and maximize efficiency from a human-focused standpoint, a CPHS–IoT system's cognitive-human layer models, comprehends, and reacts to the cognitive processes of humans, such as thinking, perception, purpose, and decision-making. The detection-human framework uses environmental sensors and mobile detectors to gather situational, cognitive, and physiological data. Cognitive processing uses AI, signal processing, and deep ML algorithms such as a convolutional neural network (CNN) to read raw information and comprehend human intent, sentiment, or tension. Using databases and reinforcement learning, preference and profile supervision combine short-term preferences with long-term characteristics to maintain dynamic user profiles. Feedback clarification uses motion recognition and natural language processing (NLP) to analyze both explicit and implicit input from users to improve system performance. Whereas adaptation and persuasion adapt system behavior according to cognitive load and user choices, decision intercession combines human purpose with cyber analytics to guarantee moral and user-friendly actions. By providing interactive visualizations and explainable AI results, the trust and awareness interface increases human confidence in the algorithm. Table 5.3 represents roles and functions with their enablers.

In a hybrid CPHS–IoT design, the cognitive-human layer is a crucial element that permits genuine interaction among sophisticated CPSs and human users. By incorporating human cognition, choices, sensations, and decision-making into the framework as a whole, it enhances the design's intelligence, adaptability, and user-centricity. The closed-loop system depicted in Figure 5.2 is

Table 5.3 Roles and functions of the cognitive-human layer.

Role	Functions	Typical methods/enablers
Detection human framework	Records situational, cognitive, and physiological information	Mobile detectors, devices that are worn, environmental microphones, and visual components
Cognitive-state approximation	Uses unprocessed signals to infer mental states such as intent, emotion, stress, attentiveness, and exhaustion	Artificial intelligence, multisensory fusion, processing of signals, and deep machine learning (medical purpose such as electrocardiogram [ECG] and CNN)
Predilection and profile supervision	Preserves changing customer models, which combine short-term desires with long-term characteristics	Profiles of users in the form of databases, information graphs, and reinforcement learning value functions
Feedback clarification	Analyses covert feedback and explicit input to improve system characteristics	NLP, reverse reinforcement learning, and motion recognition
Decision intercession	Integrates cyber analytics and human intent to generate activities that uphold user delight, protection, and ethics	Mixed initiative planners, explainable AI modules, and shared control algorithms
Adaptation and persuasion	Adapts actuation rage, warning timing, and user interface complexities to cognitive load and choices; may encourage behavior for security or effectiveness	Algorithmic influence designs, adaptive user interfaces, and attention-aware preparation
Trust and transparency interface	Builds calibrated trust by providing options, clarifications, and trust levels; it also records decisions for audits	Statistical charts, chat agents, and explainable artificial intelligence (XAI) displays

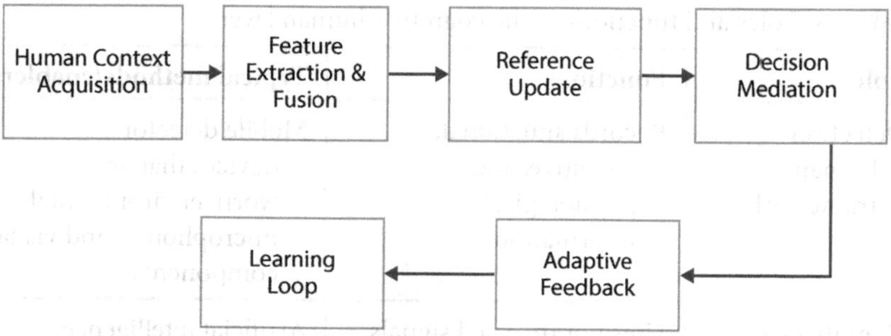

Figure 5.2 Working mechanism of cognitive-human layer.

probably utilized for adaptive human–computer interaction or context-aware decision-making. The first step is human context acquisition, which collects information about the person and their surroundings. Feature extraction and fusion, which process and combine pertinent data, come next. After then, the system goes through a reference update phase when it modifies its internal model or objectives in light of the new situation. These data are then fed into the decision mediation block, which creates an output or action. To close the loop and allow for ongoing adaptation and improvement of the decision-making process, the system includes an adaptive feedback mechanism to evaluate the decision's outcome. This feedback is then utilized by the learning loop to enhance future performance.

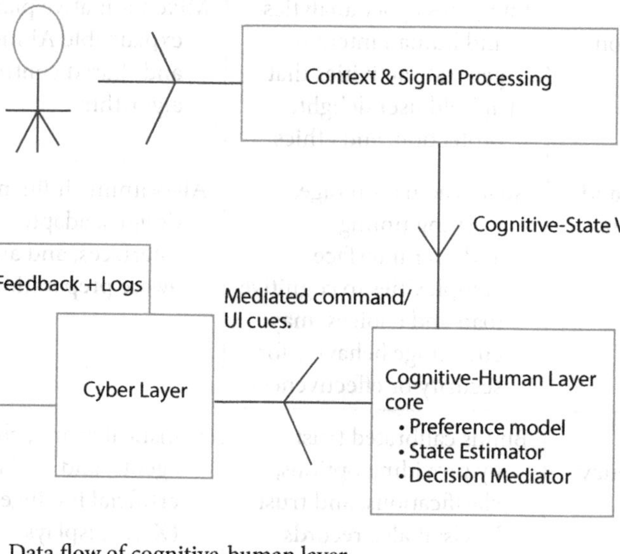

Figure 5.3 Data flow of cognitive-human layer.

A model for a cognitive-human layer system that interprets user state and mediates interactions is shown in Figure 5.3. The context and signal processing block records the user's input and contextual information first. The cognitive-human layer core receives the cognitive-state vector, which is produced by this processing and represents the user's present affective or cognitive state. The state vector is utilized by this core, which consists of a preference model, state estimator, and decision mediator, to identify the proper response. The cyber layer, which stands in for the underlying system or technology, receives a mediated command or user interface cues from the core. Lastly, a closed-loop adaptive interaction system may be created by using input + logs from the cyber layer to improve the platform's models.

5.3.2 Important Patterns of Execution

Edge-aided cognition: To protect privacy, compact devices use gadgets or the edge and only upload anonymized embeddings to the cloud when required. Dual-power adjudication: the amount of authority that AI and humans possess is determined by the blend ratio: $\alpha(t) = f(\text{confidence AI, user state, criticality})$.

Forecast switch: This provides an explanation immediately once the user delays or the prediction variability exceeds the threshold value. By analyzing human inputs and carrying out adaptive layer actions, the given Table 5.4 demonstrates how a cognitive-human layer strategy may be applied across multiple domains to improve system interaction and safety.

Table 5.4 Domain and actions of cognitive-human layer.

Domain	Human inputs	Cognitive-human layer action
Smart manufacturing	When workers left the reading of heart rate	Warning information, slow the conveyors, stimulating the suggestions
Smart grid	Domestic prevails DR event	Apprises comfort–cost curve; advises substitute valid space
Assistive driving	Motorist sleepy EEG design	Shifts to lane-cantering manner, generate gentle warning, providing rest or refreshment locations

For example, in smart manufacturing, a worrying decrease in a worker's heart rate prompts proactive actions such as warning, reducing conveyor speed, and making encouraging recommendations. When a residential demand response (DR) event takes place, a smart grid's system offers a comfort–cost curve analysis and recommends other appropriate locations for energy management. A proactive and safety-critical adaptive reaction is demonstrated in assistive driving when an electroencephalogram (EEG) pattern identifies a drowsy driver and triggers quick, layered interventions, such as putting the vehicle in lane-centering mode, issuing a polite warning, and recommending rest sites in the area.

5.3.3 Cyber Layer in Hybrid CPHS–IoT Architecture

The cyber layer serves as the system's "intelligence or brain" and is responsible for processing, data analysis, generating information, ML speculation, and decision-making. By using computations, simulations, and strategies, it turns unprocessed data into information that can be put to use. The cyber layer is the primary intellectual and connectivity foundation of a hybrid CPHS–IoT configuration. It should interpret, analyze, and maintain information collected by sensory and tangible devices. To ensure secure data transit among the physical and human components, this level makes use of robust network and cloud environments. It also makes use of machines, neural networks, and ML to generate meaningful data. Through the facilitation of robotics and rapid decision-making, the cyber layer improves system flexibility and awareness. It is crucial for connecting digital intelligence and the real world in a CPHS–IoT system.

As explained in Table 5.5, a multistep workflow for processing information and making choices is a hallmark of a strong cognitive-CPS. Gathering information *via* multiple sources is followed by data preparation, which removes contaminants and arranges the data. Context modeling subsequently applies geographical and semantic labels to the data, enabling useful analysis. The algorithm then does real-time analytics using statistical logic or rules. For more intricate anomaly recognition and forecasting, ML reasoning may have to be included. A crucial decision optimization stage solves multiobjective, constrained problems to determine the optimal course of action. Lastly, the system performs feedback to cognitive and physical layers by providing the necessary warnings, predictions, or control commands such as altering machine motions or requesting human confirmation.

ML techniques can be used in the cyber layer in hybrid CPHS–IoT. Tactical goals, specific use cases, and techniques for addressing significant problems in complicated structures utilizing different ML approaches

Table 5.5 Working mechanism of cyber layer in hybrid CPHS–IoT architecture.

Step	Function	Details
1. Data aggregation	Gather information streams from network records, physical equipment, and input from people	Multisource: network monitoring, gadgets, sensors, and app activities
2. Data preprocessing	Prepare data for analysis by cleaning, normalizing, and structuring it	Time-sync among sensors, noise mitigation, and missing value restoration
3. Context modeling	Add temporal, geographical, or semantic tags to raw data to enhance it	Zone A temperature at 3:45 pm while the machine was not in use; raise alert if tremor exceeds the limit and shift, is night
4. Real-time analytics/rule evaluation	Use statistical or rule-based logic	If vibration > threshold and shift = night, raise warning
5. Machine learning inference	Use trained models for anomaly detection, categorization, and prediction	Cloud inference, edge-ML, ensemble models, and deep learning CNN (e.g., [CNNs], long short-term memory [LSTMs])
6 Decision optimization	Solve constraints-based issues with multiple objectives	Using optimization solvers for scheduling, routing, and resource allocation (e.g., mixed integer linear programming [MILP], heuristic search)
8. Feedback to cognitive and physical layers	Send warnings, forecasts or control commands	Minimize machine speeds, remind controller, request human confirmation

are displayed in Table 5.6. For forecasting purposes such as anticipating usage of energy or breakdowns of equipment, time-series algorithms such as long short-term memory (LSTM) and autoregressive integrated moving average (ARIMA) are used. For difficulties with classification such as

Table 5.6 Types of machine learning algorithm for cyber layer.

Work	Instances	Algorithms
Prediction	Apparatus disaster, energy custom, healthiness weakening	LSTM, gradient boosting, autoregressive integrated moving average (ARIMA)
Classification	Action appreciation, danger stages, fault kinds	Support vector machine (SVM), random forest, CNN
Anomaly detection	Sensor meddling, uncommon behavior	Autoencoders, isolation forest
Reinforcement learning	Best switch approach	Q-learning, deep deterministic policy gradient (DDPG)

identifying actions or kinds of defect detection, for instance, the system uses dependable methods such as support vector machine (SVM), random forest, and CNNs. Finding anomalies is a crucial step in detecting unexpected activity or sensor manipulation, and is frequently performed by techniques such as autoencoders and isolation forest. At the very least, consecutive decisions, such as selecting the optimal transition technique, are optimized using reinforcement learning approaches such as Q-learning and deep deterministic policy gradient (DDPG). This allows the structure to absorb ideal strategies over experimentation.

In a complicated framework, probably a CPS, the cyber layer acts as the focal point for data analysis and decision-making, as seen in Figure 5.4. The cyber layer receives data from a variety of different things such as gadgets, individuals, and devices. It then sends these unprocessed data

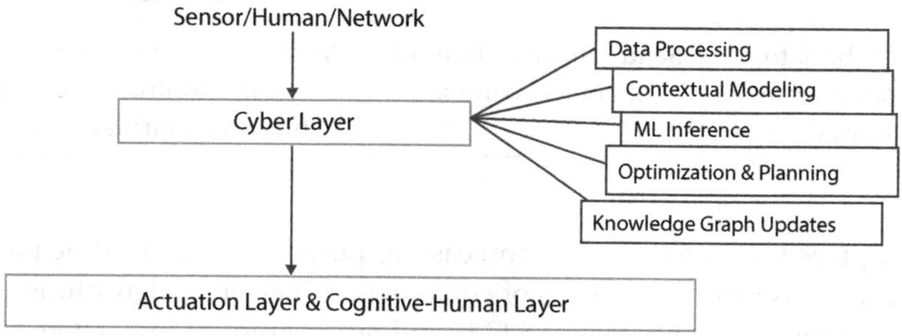

Figure 5.4 Data flow of cyber layer.

to various surveillance and analysis applications, including contextual simulation, data analysis, ML inference for classification and estimation, knowledge graph updates for continuous learning and preservation, and efficiency and scheduling. The actuator layer (mechanical control) and the cognitive-human layer (human contact and intellectual support) are then powered by the outcomes generated by these processes, thereby closing the control and intelligence loop.

Based on their position type, Table 5.7 splits data processing and storage into three balancing paradigms: cloud computing, fog computing, and edge computing. To provide quick, real-time answers and lower latency, edge computing uses lightweight foundations such as TensorFlow Lite to process data immediately on local gadgets or gateways (on entryways or edge equipment). As a middle layer, fog computing bridges the distinction between the edge networks and the main cloud by distributing compute *via* transitional nodes close to the source. It is frequently coordinated with technologies such as Kubernetes on microservers. Utilizing enormous server farms (centralized servers) and potent platforms such as AWS/GCP/Azure and Apache Spark for robust storage, intricate analytics, and worldwide coordination, cloud computing continues to be the centralized backbone.

The cyber layer supports many domains such as smart manufacturing, smart grid, healthcare, and smart agriculture. Through advanced analytical capabilities, the cyber layer plays a crucial role in promoting intelligent decision-making across a number of crucial domains, as Table 5.8 illustrates. To predict engine ruin and enable proactive maintenance, the cyber layer in smart manufacturing uses algorithms such as LSTM to enhance the preservation agenda. Estimating peak demand, identifying energy theft,

Table 5.7 Deployment types of the cyber layer.

Type	Occurring places	Tools
Edge computing	On entryways or edge equipment	TensorFlow Lite, Edge Impulse, Nvidia Jetson
Fog computing	Transitional nodes near to source	Kubernetes on micro servers
Cloud computing	Centralized servers	AWS/GCP/Azure, Apache Spark, Databricks

Table 5.8 Cyber layer domain and its functions.

Domain	Cyber layer function
Smart manufacturing	Forecast engine ruin with LSTM; enhance preservation agenda
Smart grid	Estimate highest request, sense stealing, notice DERs optimally
Healthcare	Conclude patient worsening from multimodal vitals by means of ML
Smart agriculture	Analyze climate–soil–crop statistics for irrigation timing by means of decision trees

and managing distributed energy resources (DERs) optimally are just a few of the smart grid's security-focused and optimization-focused features. To combine multimodal vital signs and determine patient deterioration for early intervention, the cyber layer in healthcare uses ML. Lastly, decision trees are used in smart agriculture to evaluate intricate crop, soil, and climatic data to optimize critical processes such as irrigation scheduling.

(i) Important Cyber Layer Features

As the computational part of the brain that links physical perception to actionable results and incorporates the human aspect, the cyber layer serves as the intellectual core of a CPHS. Achieving hybrid intelligence through effective data and control management is at the heart of its most important aspects. This incorporates real-time data processing and analytics, which instantly transforms unprocessed sensor data into insights that can be used to make decisions. The system's fundamental logic is carried out by sophisticated intelligent control algorithms and AI/ML models. The other features are listed below.

- **Adaptable**: Able to manage substantial amounts of diverse data.
- **Intelligent**: Incorporates learning and adapting AI/ML models.
- **Immediate capable**: Facilitates low-latency decision-making and inference.
- **Interoperable**: Capable of communicating with network, physical, and cognitive-human layers.
- **Trustworthy**: Able to incorporate audit trails and explainable AI modules.

5.4 Physical Layer in Hybrid CPHS–IoT Architecture

The bottommost layer of the CPHS–IoT stack is called the physical layer, and it is made up of embedded hardware, electrical and mechanical systems, sensors, actuators, and other components that have direct physical interactions. It makes interaction with the environment, actuation, and sensing possible. Table 5.9 explains the various stages of the physical layer and their functions.

5.4.1 Step-by-Step Workflow of the Physical Layer

Step 1: Interaction with the environment: A sensor, such as an accelerometer, may identify a machine's vibration. A transducer is a device that transforms physical impulses, such as pressure, into electrical ones.
Step 2: Signal conditioning: This step utilizes an analog-to-digital converter (ADC) to convert data to a digital format, eliminate noise, and

Table 5.9 Various stages of physical layer and their functions.

Stage	Function	Description
1. Sensing	Gathers physical/ environmental constraints	Routines devices to notice motion, heat, vibration, compression, heartbeat rate, etc.
2. Signal conditioning	Makes raw signals for processing	Filtering, amplification, digitization *via* ADC (analog-to-digital conversion)
3. Data transmission	Directs information to higher layers	Utilizes protocols such as Zigbee, Wi-Fi, BLE, LoRaWAN, MQTT
4. Actuation	Replies to procedures from cyber/cognitive layers	Controls motors, valves, lights, relays, alarms, etc.
5. Feedback loop	Terminates the command loop after sensing output effects	Verifies the successful execution of a command (such as rotation or switch-on)

amplify weak signals. A signal is sent by a microcontroller to a communication interface (e.g., an ESP32 sent *via* Wi-Fi).
Step 3: Actuation: The cyber layer determines that the motor must stop and instructs the actuator accordingly. To change states, an actuator (such as a servo) receives a pulse-width modulation (PWM) signal.
Step 4: Response: The location sensor communicates the status back upstream after confirming that the motor has stopped. The physical layer consists of various components such as sensors, actuators, transducers, and microcontrollers/embedded devices. Table 5.10 explains the various components and their functions.

By describing the physical hardware parts and their direct activities in context, Table 5.11 illustrates how the physical layer serves as the basis for different automated systems. Robotic arms and vibration sensors are crucial parts of a smart factory for functions such as fault detection and accurate robotic control. Electrocardiogram (ECG) electrodes and infusion devices are examples of technologies used in medicine that carry out crucial tasks including assessing patient vitals and modifying medicine circulation. In a similar vein, smart homes use gadgets such as motion detectors and heaters to regulate protection and climate control. The physical layer acts as the connection between technology and the physical world, facilitating data collection and the execution of control commands in all domains, including agriculture with its soil moisture sensors and irrigation pumps and the smart grid with its smart meters and breakers.

Table 5.10 Components and their functions of physical layer.

Component	Role
Sensors	Gather atmospheric and physical data, such as ECG, pressure, and temperature
Actuators	Execute operations (such as rotating, opening/closing, and starting/stopping)
Transducers	Change the form of energy (e.g., piezoelectric for force-to-electric)
Microcontrollers/embedded devices	Connection among the communication layer and sensors/actuators
Power supply units	Make sure the components have steady supply
Wired/wireless interfaces	Allow data to move to the network layer

Table 5.11 Domains of physical layer and components.

Domain	Physical layer components	Actions
Smart factory	Vibration sensors, robotic arms, temperature probes	Detect faults, control robotic precision
Healthcare	ECG electrodes, infusion pumps, wearable biosensors	Monitor vitals, adjust medication flow
Smart home	Motion sensors, smart locks, thermostats	Detect movement, control temperature, lock doors
Smart grid	Smart meters, voltage sensors, breakers	Measure usage, control power delivery
Agriculture	Soil moisture sensors, irrigation pumps	Monitor soil, trigger watering system

A typical CPS layered architecture is shown in Figure 5.5, starting with the physical layer. This layer includes built-in controllers to operate them and communicate with the surroundings right away, sensors to collect data, and actuators to carry out physical operations. The network and communication

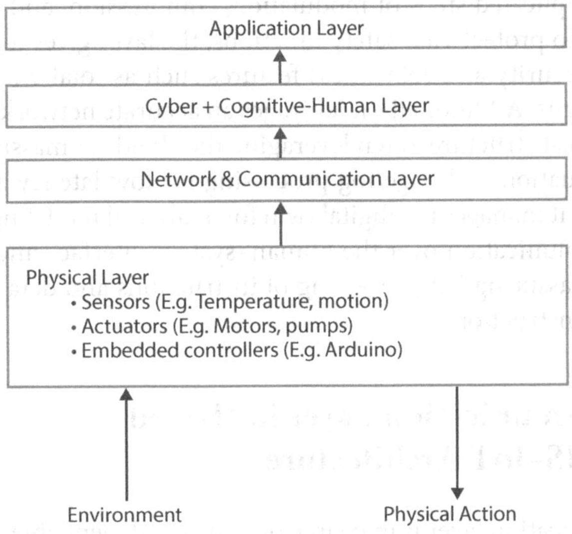

Figure 5.5 Work flow of the physical layer.

layer, which manages safe and dependable data transfer, receives data from the physical layer and passes them upward. The cyber + cognitive-human layer is where intelligence (processing, decision-making, and user engagement) is carried out after these data have arrived. At the top is the application layer, which offers the user contact and particular domain services. To start a physical action, feedback from the upper levels travels down to the physical layer.

5.4.2 Important Characteristics of the Physical Layer

Direct communication with the physical environment is the cornerstone of the CPHS–IoT stack. The system must be capable of a real-time response to instructions or modifications. Reliability is crucial because malfunctions in sensors or actuators can impact the system as a whole. These components are extremely varied, ranging from industrial robotics to healthcare sensors. Being the interface between the digital and physical worlds is one of the physical layer's primary functions in CPHS that are connected to the IoT. To enable the crucial closed-loop control, it mainly contains sensors that decode physical features from the real world into numerical signs and actuators that interpret digital instructions into physical movements. This layer must manage a wide range of diverse communication mediums while upholding strict real-time harmonization for critical control processes, including high-speed cable links to low-power wireless networks for distant IoT gadgets. To maintain information accuracy and confidentiality in the midst of external interference and constraints, it also regulates the complicated steps of modulation, compression, and information translation. To protect vital safety functions, the layer gives top priority to strict cybersecurity and robustness features such as total encryption and resilient designs. Additionally, it leverages an accurate network monitoring and data transit structure, often leveraging the cloud for massive data storage and evaluation and edge/fog processing for low-latency applications. Prominently, it manages the digital twin for analytical modeling and allows human communication over the human–system interface and communication logic, assuring flat processing of instructions and data flow during the CPHS construction.

5.5 Communication Layer in Hybrid CPHS–IoT Architecture

The communication layer is in charge of ensuring dependable and efficient data transfer among various actuators, sensors, devices at the edge, services in the cloud, and user experiences. With its integration of the physical, cyber, and cognitive-human layers, it enables seamless communication across the framework. The communication layer's operation is described in Table 5.12.

Table 5.12 Working mechanism of the communication layer.

Stage	Function	Description
1. Data capture from source	Gather information from wearables, detectors, or peripheral devices	Sensor data, either raw or preprocessed and reaches *via* the communication channel
2. Protocol selection and formatting	Uses standard protocols for communication to encrypt data	Depending on the requirements for speed, power, and range, MQTT, CoAP, HTTP, Zigbee, and 5G
3. Routing and addressing	Establishes the network's data flow and guarantees accurate delivery	Use mesh node IDs, medium control access MAC IDs, or IP addresses to direct packets
4. Transmission medium	Sends data *via* wired or wireless networks	Bluetooth, satellite, Ethernet, Wi-Fi, cellular, and LoRa
5. QoS and reliability checks	Checks for data integrity, and gives priority to important packets	Retries, encryption, buffering, traffic shaping, and acknowledgements
6. Reception at destination	Data is delivered to cloud, edge, or gateway systems for examination	Data are decoded, verified, and stored or transmitted to the cyber layer *via* receivers
7. Feedback/control data flow	Transmits commands to actuators or users from the cyber layer	The dashboard updates, smartphone notifications, and actuator inputs

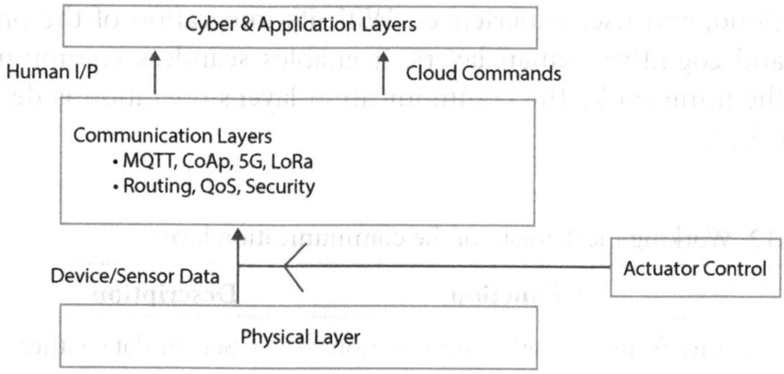

Figure 5.6 Connection of communication layer.

The communication layer uses various protocols such as Zigbee, Bluetooth low energy (BLE), Z-Wave, Modbus, Internet protocol version 6 (IPv6), IPv6 over low-power wireless personal area networks (6LoWPAN), long-range wide-area network (LoRaWAN), narrowband IoT (NB-IoT), transmission control protocol (TCP), user datagram protocol (UDP), quick UDP Internet connections (QUIC), MQTT, CoAP, HTTP, and WebSockets to communicate with other layers such as the device layer, network layer, transport layer, and application layer. Figure 5.6 explains the connection with various layers of communication within the layer.

5.5.1 End-to-End Communication Flow Example for Smart Manufacturing

A computer numerical control machine's tremor sensor detects unusual activity. An embedded board receives a vibration signal from the sensor *via* universal asynchronous receiver–transmitter. An edge port receives data from a Wi-Fi router *via* an Internet protocol (IP) network. The edge gateway uses HTTPS to transport data to the cloud server. The cloud server delivers back an authorization message after conducting analysis. To halt the machine, an instruction is sent *via* LoRaWAN to the actuator. The provider of services receives real-time updates through a mobile app that uses WebSockets.

5.6 Sensor Layer in CPHS–IoT Architecture

Mechanism of operation: In a CPHS–IoT design, the sensor layer is the base tier. Its main function is to gather data in real time from systems and the physical world, which is essential for facilitating automated processes, cognition feedback, and more advanced decision-making.

5.6.1 Operating Mechanism

Detection and gathering data: A variety of sensors, including movement, temperature, pressure, proximity, humidity, gas, biosensors, and others, are incorporated in systems or placed throughout the environment. These sensors translate physical variables into digital data or electrical impulses while constantly tracking them.

Signal conditioning: Unreliable or noisy raw sensor outputs are common. The data are processed by circuits such as filters, amplifiers, etc. to ensure accuracy and usability.

Optional internal processing: To reduce the communication burden, certain advanced sensors use microcontrollers or edge computing units to carry out initial information filtration, detection of anomalies, or grouping directly.

Communication with upper layers: Wireless (Wi-Fi, Zigbee, BLE, LoRa) or cable (ethernet, can) protocols are utilized to direct the processed sensor data to the communication layer. based on the application, the information is subsequently sent to either the cognitive-human layer for feedback or the cyber layer for analysis.

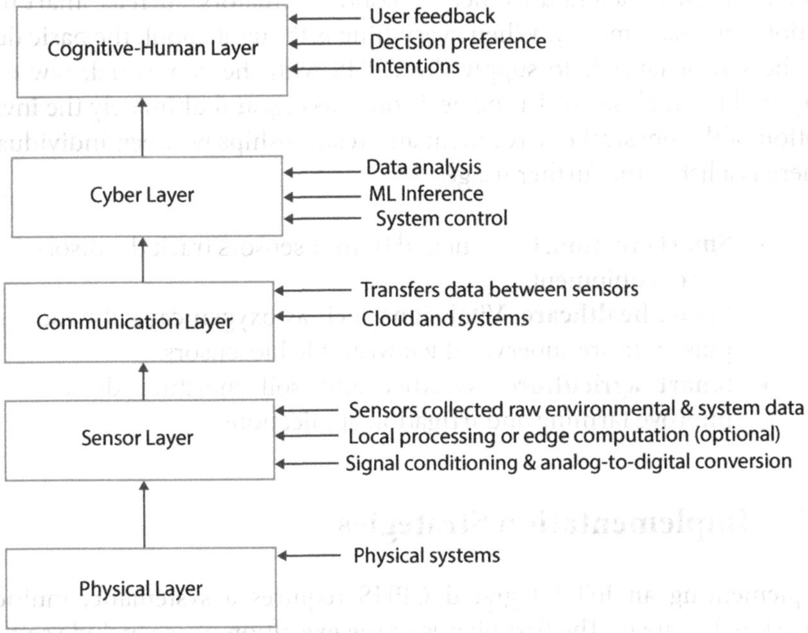

Figure 5.7 Communication of CPHS–IoT layers.

Real-time monitoring: To maintain the system's responsiveness and context awareness, the sensor layer makes sure that changes are made in real time or almost real time. Figure 5.7 shows the various layers of communication within CPHS–IoT along with their functions.

5.6.2 Examples of Uses

As the "eyes and ears" of a CPHS, the sensor layer (also called the perception layer or device layer) continuously gathers real-time data and converts them into digital signals, acting as an interface between the digital and real scenarios. Utilizing wearable sensors, this layer of smart healthcare continuously transmits data for distant patient surveillance and prompt emergency intervention by tracking a patient's important data such as heart rate, blood pressure, and oxygen saturation. Whereas proximity sensors improve worker safety by tracking their location in relation to robots, the smart manufacturing (Industry 4.0) layer uses vibration, temperature, and acoustic sensors mounted on machinery to identify anomalies, enabling predictive maintenance and averting expensive equipment failures. Furthermore, smart agriculture uses specific sensors such as moisture levels in the soil, pH, and chemical monitors to gather information on the environment to optimize yields of crops and preserve resources. After then, the cyber layer receives these data and uses them to control actuators, such as smart irrigation devices, remotely. When everything is thought about, the basic duty of the sensor layer is to supply the CPHS with the contextual, raw data required for analysis, real-time decision-making, and ultimately the invigilation of the physical environment and relationships between individuals. There is a list of the further usage.

- **Smart creation**: Current and tremor sensors track the disorder of equipment.
- **Smart healthcare**: Vital signs such as oxygen capacity and pulse rate are supervised *via* wearable biosensors.
- **Smart agriculture**: Weather and soil moisture devices improve farming and irrigation applications.

5.7 Implementation Strategies

Implementing an IoT-integrated CPHS requires a systematic, multidimensional strategy. The first phases in the execution phase include setting clear business objectives and conducting small-scale trials to evaluate and enhance the system before a large-scale rollout. Its architecture needs a hierarchical structure that particularly incorporates the human element to enable real-time data analysis and low-latency selection, which are critical for system security. It uses an edge/fog processing approach to accomplish this. Overall, the architecture must prioritize portability by utilizing defined interfaces to link different devices. The system's cognition is driven by AI/ML models enabling sophisticated analytics, including prediction and human activity recognition. Finally, an effective security-by-design method, including robust authorization processes and network division, in addition to extensive instruction and change management, must be prioritized to ensure satisfaction among users and operational durability when faced with combined cyber, physical, and psychological risks.

5.7.1 Design Principles

The specified strategies should be followed by an effective CPHS–IoT system. The effectiveness of a CPHS–IoTCPHS–IoT system depends on several important guidelines. The arrangement should follow the design-for-security rules, implementing robust control of entry and data protection across all physical and cyber elements to handle the increased threat field. Transparency and uniformity are crucial to ensuring uninterrupted interaction among various IoT gadgets, past technologies, and the cyber framework. The platform has to be human-centric, featuring simple user interfaces to continuous, relevant data, to ensure acceptance by users and facilitate human decision-making. An alternative computing strategy is required for managing swift control loops and important security measures so that immediate processing will not be hindered.

Furthermore, an ongoing surveillance and maintenance strategy that integrates automatic software upgrades and predictive analytics (AI/ML) is essential for both efficiency improvement and defense against malfunctions and evolving threats. Finally, the structure must be flexible to accommodate future growth and ethical requirements while safeguarding the management of data including user confidentiality. The following is a list of additional recommendations:

- **Flexibility**: Improves maintenance by decoupling system functions.
- **Resilience**: Uses cloud–edge architectures to control data growth.
- **Privacy**: Makes the most of multilayer secure access control and detection of intrusions.
- **Interconnection**: Makes integration easier using free standards and APIs.

Table 5.13 Layers and protocols.

Layer	Protocols
Device	Zigbee, BLE, RFID
Network	MQTT, CoAP, 6LoWPAN
Application	REST, OPC-UA
Security	OAuth 2.0, TLS, DTLS

5.7.2 Protocols and Standards

The proper operation of the IoT and CPHS architecture depends on the subsequent protocols that enable interaction across levels. Device-layer techniques such as RFID, BLE, and Zigbee provide low-power, short-range interaction among detectors and motors. The network layer controls data transportation using compact machine-to-machine protocols such as MQTT, CoAP, and 6LoWPAN, which are meant for IoT gadgets with limited resources. The application layer communicates with software applications using standards such as REST and open platform communications unified architecture (OPC-UA) to facilitate sharing of information, instruction performance, and corporate system connectivity. Last but not least, solid safety is handled by techniques such as OAuth 2.0 for authentication and transport layer security (TLS) or datagram-TLS (DTLS) for ensuring the safety and security of the information sent across the network and application levels. Specifications of the layer-wise procedures are explained in Table 5.13.

5.8 Case Study of Smart Agriculture Using CPHS–IoT System

5.8.1 Objective

The primary objective of CPHS–IoT system adoption in agricultural sectors is to achieve high-yield, profitable agriculture by creating a closed-loop, cognitive ecosystem. This system aims to improve the effectiveness of resources, including reducing the use of fluids and chemical fertilizers by tracking the ground, weather, and crop state using current information through IoT sensors. Among the primary goals is to enhance human judgment by providing farmers with modeling and practical data for activities such as smart watering and pest surveillance. Further, the system seeks to reduce worker labor and human mistakes while enhancing productivity by automating physical tasks with sensors.

The ultimate goal is to use robust and adaptable cyber–physical–human collaboration to guarantee better quality output, less environmental effect, and increased economic viability for agricultural enterprises. The goal is to use a CPHS–IoT system to integrate sensors, automation, AI, and farmer input to maximize agricultural productivity and resource utilization on a farm.

5.8.2 Layer-Wise Implementation

(i) Physical layer: Components: Irrigation pumps, valves, tractors, drones, automated sprinklers, crops, soil, and weather environment.
Function: Creates the actual farming environment that communicates with actuators and sensors.
(ii) Sensor layer: Components: Soil moisture sensors, temperature and humidity sensors, pH sensors, light sensors (for photosynthesis monitoring), weather stations, and camera drones (for crop surveillance).
Function: Keep an eye on the soil and atmospheric conditions at all times. Compile up-to-date information on field characteristics and the health of crops.
(iii) Communication layer: Components: Wireless sensor networks (LoRa, Zigbee, Wi-Fi), gateways, and edge devices.
Function: Send gathered sensor data to local or cloud servers. Make it possible for actuators (such as irrigation systems) to receive remote control signals. Make sure all subsystems can communicate securely and with minimal latency.
(iv) Cyber layer: Components: Cloud-based analytics platform, AI/ML algorithms, decision-making dashboards, and crop simulation models.
Function: Examine information to find trends such as nutritional deficits, irrigation requirements, or infestations of pests. Estimate yields and apply fertilizer and pesticides as efficiently as possible. Offer suggestions or automatic processes (e.g., if the soil is dry, begin drip irrigation).
(v) Cognitive-human layer: Components: Farmer's mobile app/dashboard, alerts and recommendation systems, and feedback modules.
Function: Enables farmers to enter their choices (e.g., irrigation schedule, crop kind). AI makes advice to farmers, who can override automation if necessary.
Training loop: The system iterates choices in response to human input and results.

5.8.3 Smart Agriculture CPHS–IoT System

Farm oversight and ecological sustainability have significantly improved as a result of a smart agriculture CPHS–IoT system. Extremely accurate, real-time interventions based on ML analysis of environmental data result in an average 15%–20% increase in crop yields. Automated variable-rate irrigation that is directly linked to ground humidity can cut water use by up to 30%–40%, achieving the goal of resource optimization. Additionally, the method effectively minimizes environmental runoff and lowers input costs by reducing the application of pesticides and fertilizers by 10%–25%. According to human-centric statistics, farmers spend 40% less time on manual field scouting when they are guided by the system's digital twin visualizations and predictive alarms. All things considered, combining the cyber, physical, and human components results in a 25% reduction in operating energy expenses and creates a strong, resilient agricultural operation with improved data-driven decision-making skills. Water usage was decreased by using intelligent irrigation. Higher yield is a result of prompt nourishment control and pest identification. Improved producer decision-making is through recommendations supported by AI. Labor reductions are through distant surveillance and automation.

5.9 Advantages and Challenges

In the given case study, because technology, biology, and human factors are closely intertwined, the CPHS approach to smart agriculture, when combined with IoT, offers substantial benefits while also posing difficult problems.

5.9.1 Advantages

By combining the IoT with CPHS, significant benefits are produced, including extremely responsive, intelligent, and adaptive systems. Through the provision of real-time, context-aware information that enables efficient choice-making, these tools transform individuals into smart managers. They significantly boost manufacturing and productivity by automating monotonous physical tasks with actuators and mechanical elements. CPHS–IoT reduces waste and operational costs while enhancing resource efficiency by facilitating precise, site-specific administration. Furthermore, problem diagnosis and resilience to damage are improved by the predictive analytics and constant tracking capabilities, which provide quick disaster recovery and minimize downtime. Ultimately, CPHS–IoT establishes a win–win collaboration between technology and human resources, yielding superior, high-volume results in a range of complex domains, such as smart agriculture, manufacturing in industrial settings, and healthcare. Human-centric cognition incorporates people's beliefs, actions, and remarks.

Improved trustworthiness: Adaptive control reduces system breakdowns. **Better decision-making**: Combines data analytics and intuition. Constant monitoring across distributed systems is essential for seamless monitoring and mitigation.

5.9.2 CPHS–IoT Challenges

The exertion of thoroughly connecting the three domains is the central hindrance to the positioning of CPHS shared with IoT. The extensive network of IoT devices significantly expands the attack surface, raising the possibility of data theft or unauthorized physical manipulation, making cybersecurity a top issue. A technical difficulty is interoperability and heterogeneity, which call for smooth communication between various vendor-specific devices, legacy equipment, and central platforms. Network capacity is strained and precise, context-aware AI models are required due to the vast volume and diversity of data collected, which calls for sophisticated real-time data management and processing at the edge. Human-related issues also include the difficulties of creating interfaces that really enhance human jobs rather than make them more difficult, resistance to change, and a possible lack of technical proficiency among users. Lastly, widespread adoption may be hampered by the initial high cost and moral dilemmas surrounding data privacy and surveillance. Table 5.14 explains the CPHS–IoT challenges in each area and their descriptions.

Table 5.14 CPHS–IoT challenges.

Challenge	Description
Data privacy	Ensuring secure storage and transmission of data
Latency	Achieving real-time performance in networked systems
Model complexity	Managing hybrid dynamics and human variability
Trust in AI	Designing systems humans can understand and rely on

5.10 Conclusion

Management and engineering technologies are being revolutionized by the hybrid CPHS–IoT concept. It makes it easier for systems to transition from strict, rule-based management to ones that are smart, flexible, and sensitive to human needs. The combined approach facilitates the development of scalable, reliable, and adaptable structures that are necessary for future technology environments by utilizing the advantages of the two models. Multidisciplinary study as well as effective execution strategies will be crucial as this field develops to solve unresolved issues and realize the complete potential of CPHS–IoT in usage. Virtual twin collaboration involves simulated and monitored dynamic system copies, whereas federated instruction enables training models on dispersed nodes without exchanging data. Bio-cyber connections integrate brain–computer interfaces with metabolic detectors, and quantum IoT leverages quantum technology to analyze data quickly and securely. Additionally, self-adaptive mechanisms function as independent entities that modify their configuration in response to objectives and circumstances.

AI disclosure statement: The author confirms that no artificial intelligence (AI)-generated content or AI-assisted writing tools were used in the preparation of this chapter (including summaries and the abstract). All text, analysis, and conclusions are the result of the author's original work.

References

1. Lee, E.A., Cyber physical systems: Design challenges, in *Proc. 11th IEEE Int. Symp. Object and Component-oriented Real-time Distributed Computing (ISORC)*, 11, 363, 2008.
2. Lee, E.A., Cyber-physical Systems - Are computing foundations adequate, in: *Position Paper for NSF Workshop on Cyber-Physical Systems: Research Motivation, Techniques and Roadmap*, 2, 9, 2006.
3. Calder, M., Craig, C., Culley, D., De Cani, R., Donnelly, C.A., Douglas, R., Edmonds, B., *et al.*, Computational modelling for decision-making: where, why, what, who and how. *R. Soc. Open Sci.*, 5, 172096, 2018.
4. Chadwick, C., Betzig, S., Hu, F., Cyber-Physical Systems Concepts, in: *Cyber-Physical Systems: Integrated Computing and Engineering Design*, 2, 134, CRC Press (Taylor & Francis Group), Boca Raton, FL, September 18, 2013.
5. Kotronis, C., Nikolaidou, M., Tsadimas, A., Kapos, G.-D., Dalakas, V., Routis, I., Bardaki, C., Dimitrakopoulos, G., Anagnostopoulos, D., Engaging human participants in CPHS design: A multi-view approach transforming

human criticalities to system design requirements. *IEEE Internet Things J.*, 30, 101443, 2025.

6. Rajkumar, R., Lee, I., Sha, L., Stankovic, J., Cyber-Physical Systems: the next computing revolution, in: *Proc. 47th Design Automation Conference,* 12, 736, 2010.
7. Nimbe, P., Weyori, B.A., Adekoya, A.F., Models in quantum computing: a systematic review. *Quantum Inf. Process.*, 20, 80, 2021.
8. Silva, G.C. and Kaminski, P.C., From Embedded Systems (ES) to Cyber-Physical Systems (CPS): An Analysis of Transitory Stage of Automotive Manufacturing in the Industry 4.0 Scenario. *SAE Tech.*, 36, 230, 2016.
9. Koc, H., Shaik, S.S., Madupu, P.P., Reliability modeling and analysis for cyber physical systems, in: *Proc. IEEE 9th Annual Computing and Communication Workshop and Conference (CCWC)*, 9, 448, 2019.
10. Miclea, L. and Sanislav, T., About dependability in cyber-physical systems, in: *Proc. 9th East-West Design & Test Symposium (EWDTS)*, 9, 17, 2011.
11. Tortorella, M., *Reliability, Maintainability, and Supportability: Best Practices for Systems Engineers*, John Wiley & Sons, 5, 127, United States, 2015.
12. Goknil, A., Nguyen, P., Sen, S., Politaki, D., Niavis, H., Pedersen, K.J., Suyuthi, A., Anand, A., Ziegenbein, A., A systematic review of data quality in CPS and IoT for industry 4.0. *ACM Comput. Surv.*, 55, 38, 2023.
13. Biradar, H.B. and Shabadi, L., Review on IoT based multidisciplinary models for smart farming, in: *Proc. 2nd IEEE Int. Conf. Recent Trends in Electronics, Information & Communication Technology (RTEICT)*, 2, 1923, 2017.
14. Seshia, S.A., Hu, S., Li, W., Zhu, Q., Design automation of cyber-physical systems: Challenges, advances, and opportunities. *IEEE Trans. Comput.-Aided Des. Integr. Circuits Syst.*, 36, 1421, 2016.
15. Flammini, F., Dobrilović, D., Gaglione, A., Tokody, D., LoRaWAN Technology Mapping to Layered IoT Architecture. *AIIT Access*, 29, 148, 2020.
16. Al-Sarawi, S., Anbar, M., Alieyan, K., Alzubaidi, M., Internet of Things (IoT) communication protocols, in: *Proc. 8th Int. Conf. Information Technology (ICIT)*, 23, 685, 2017.
17. Atzori, L., Iera, A., Morabito, G., The Internet of Things: A survey. *Comput. Netw.*, 54, 2787, 2010.
18. Ngu, A.H.H., Eyitayo, J.S., Yang, G., Campbell, C., Sheng, Q.Z., Ni, J., An IoT edge computing framework using Cordova accessor host. *IEEE Internet Things J.*, 9, 671, 2021.
19. Ma, M., Lin, W., Pan, D., Lin, Y., Wang, P., Zhou, Y., Liang, X., Data and decision intelligence for human-in-the-loop cyber-physical systems: References model, recent progresses and challenges. *J. Signal Process. Syst.*, 90, 1167, 2018.
20. Jamshidi, M. (Ed.), *Systems of Systems Engineering: Principles and Applications*, CRC Press, Boca Raton, 2017.

6

Training Deep Neural Networks for Spatiotemporal Cyberphysical Human Systems

N. Sathish[1]*, R. Nagaraj[2], P. Arul[3] and Pham Chien Thang[4]

[1]*Department of Electronics and Communication Engineering, Sri Venkateswara College of Engineering, Sriperumbudur, Tamil Nadu, India*
[2]*Department of CSE (AI & ML), Sri Sairam Engineering College, Chennai, Tamil Nadu, India*
[3]*Department of Electronics and Communication Engineering, University College of Engineering, Anna University Panruti, Tamil Nadu, India*
[4]*Faculty of Journalism and Communication, TNU-University of Sciences, Thai Nguyen, Vietnam*

Abstract

Spatiotemporal cyberphysical human systems (CPHSs) are revolutionizing multiple sectors, such as smart cities, healthcare, intelligent transportation, and industrial automation through the integration of real-time sensing, computation, and human interaction. These systems provide extensive and intricate spatiotemporal data; they render deep learning methodologies, especially deep neural networks (DNNs), crucial for modeling, forecasting, and decision-making. Training DNNs for spatiotemporal complex-physical systems poses considerable hurdles, including high-dimensional data processing, dynamic environmental variations, data sparsity, and computational limitations. This chapter examines approaches for the effective training of DNNs in spatiotemporal CPHS. It examines essential methodologies including recurrent neural networks, long short-term memory, transformers, and graph neural networks for addressing spatial and temporal relationships. The chapter discusses optimization methodologies, transfer learning, federated learning, and edge artificial intelligence to enhance training efficiency, adaptability, and scalability. This chapter presents developments in DNN training

**Corresponding author*: n.satyish@gmail.com; ORCID: 0009-0000-3507-6337

Anitha Velu, Prasanth Aruchamy, Raghu Ramamoorthy, Rajesh Kumar Dhanaraj and Manjula Arunraj (eds.) Cyber Physical Human Systems: Emerging Trends and Applications, (131–152) © 2026 Scrivener Publishing LLC

specifically designed for CPHS, offering insights into the development of intelligent and resilient systems for practical applications in dynamic cyberphysical environments.

Keywords: Neural networks, optimization, cyberphysical human systems, artificial intelligence, machine learning

6.1 Foundations of Spatiotemporal Cyberphysical Human Systems

Modern systems are becoming more interactive and context-sensitive, and the rigid distinctions between the cyber world, the physical world, and the human world are exhausted. cyberphysical human systems (CPHSs) as an interdisciplinary paradigm realize a strong integration of sensing, computation, control, and human cognition. The needed dimensions are space (e.g., location, spatiotemporal relations) and time (e.g., sequence, duration, dynamics) in the spatiotemporal CPHSs, the definitions of which determine the system behavior and decisions made there. This chapter explores the historical development of CPHSs, and dwells on the relevance of human-in-the-loop dynamics, and the key role of artificial intelligence (AI) and deep learning to secure the necessary form of adaptivity, intelligence, and the resultant attainment of human-centric system behavior.

6.1.1 Evolution and Definition

CPHSs are a major extension of traditional cyberphysical systems (CPSs) and make specific attempts to incorporate humans and their behaviors and decisions into the control and computational loop. Whereas CPSs integrate computation, networking, and physical activities to allow intelligent automation, CPHSs build on this paradigm by injecting human engagement as an interactive and dynamic element into the paradigm. Such an addition makes the system a triad of cyber, physical, and human spaces. Traditionally, the development of standalone embedded systems to networked CPSs and then CPHSs is the result of the increasingly complex and interactive requirements that are demanded by current applications [1]. The human agent is not only a user in other parts of an intelligent system consisting of smart healthcare, collaborative robotics, and intelligent transportation, but the human agent is an agent that transforms, reacts, connects, and coevolves with the system.

Spatiotemporal CPHSs increase the location-sensitive and time-sensitive characteristics of data and interaction. Such systems are always aware, discover, and respond to spatial arrangements (e.g., layout of the environment, location of a user) and temporal variations (e.g., movement patterns, physiological patterns) resulting in a highly contextualized and personalized computing world.

6.1.2 Human-in-the-Loop System Dynamics

An important feature of CPHSs is that they will have human-in-the-loop feedback mechanisms. In sharp contrast to conventional automation systems, CPHSs have to come to grips with a variety of human behaviors that are potentially nondeterministic, context-intensive, and dynamically changing. The architecture of the CPHS feedback loop is illustrated in Figure 6.1.

Such systems normally work on the following loop:

- **Perception**: Sensors gather spatiotemporal information about the surroundings as well as human actors (e.g., location and gesture, voice and biometrics).
- **Computation**: The interpretation of data are the creation of an algorithmic representation of the intentions, behaviors, and states of humans and systems.

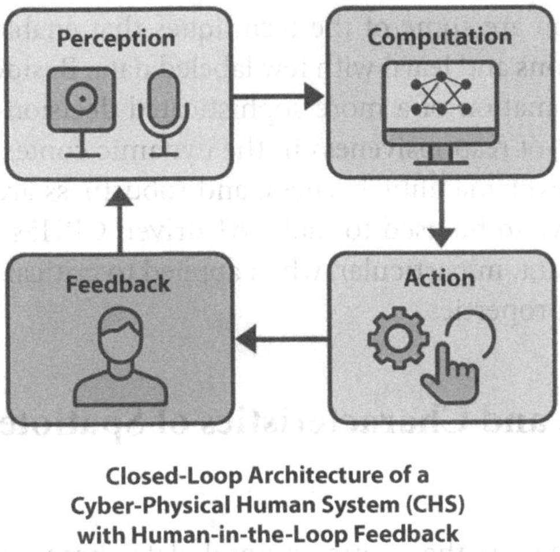

Figure 6.1 CPHS feedback loop.

- **Action**: It creates control signals to alter the behavior of the system or report back to the human.
- **Feedback**: The reaction of humans to the output of the system is observed, and it is a closed loop.

Such a dynamic interaction loop also brings forward a few challenges, including latent requirements, modeling of the user, behavior uncertainty, and adapting to the feedback. Moreover, safety and comprehensibility pose a significant problem whenever human life is a direct consequence of the decisions taken (self-driving cars, surgical robots, or assistive tools).

6.1.3 Role of AI and Deep Learning

AI, especially deep learning, has turned out to be a pillar toward the creation of intelligent spatiotemporal CPHSs [2]. Such systems have to absorb complex, dynamic, and high-dimensional data both environmentally and through human interactions. CPHSs can extract meaningful patterns out of the stream of both spatial and temporal data, making use of deep neural networks (DNNs), which include convolutional neural networks (CNNs), recurrent neural networks (RNNs), and transformers. This brings strength to systems to accomplish tasks such as human activity recognition, prediction of behavior, and context-sensitive decision-making to an amazing extent. Within the domain of human-based systems, the AI is essential in displaying purpose-defining capabilities, user behavior modeling, and real-time adaptive responses. Transfer learning, reinforcement learning, and online adaptation are some of the techniques that enable CPHSs to customize interactions and learn with few labeled data. Besides, deep learning enables the automation of a more sophisticated decision-making process and some extent of responsiveness in the dynamic contexts. Nevertheless, factors such as explainability, fairness, and robustness are central considerations that have to be used to make AI-driven CPHSs trustworthy and ethically compliant, in particular, when applied to critical procedures with safety-sensitive properties.

6.2 Nature and Characteristics of Spatiotemporal Data

CPHSs are based on the spatiotemporal data, because they track the changing state of the environment and the human being involved in it.

As opposed to their static counterparts, spatiotemporal data can and will change with time and space, creating an opportunity (or a challenge depending on the perspective) for modeling and interpretation. It is critical to know their peculiarities to develop intelligent systems that could respond adequately to the context, make decisions on time, and be adaptive to real-life variability.

6.2.1 Temporal Dynamics: Trends, Sequences, and Cycles

CPHSs measure temporal data as events and behaviors that vary in time and represent sequences, trends, and circular patterns. It is important to model the interdependencies over time regardless of whether there is a need to identify physiological signals, vehicle movements, or understand user behaviors [3]. These processes tend to be nonlinear and data-dependent in that models capable of long-range-dependence modeling and irregular sampling are required. There exist deep learning architectures such as long short-term memory (LSTM), gated recurrent units (GRUs), and temporary attention attainment to capture such complex temporal dynamics to make precise time-dependent forecasts.

6.2.2 Spatial Context: Structure, Location, and Interactions

Spatial data in CPHSs include data that fold the physical layout and mobility of objects, as well as the interactions among them, including but not limited to users, devices, and environments. Contextual spatial understanding enables the systems to read the positioning of the objects or people in real-life situations and the configurations of spatial positions and the role they play in the interaction. To give one example, in a smart environment or care monitoring, position can be used to infer intent or detect abnormalities, or trigger some context-sensitive services when one knows, for instance, the relative position to sensors or other persons. Spatial context is also used to allow the system to localize actions within the area and optimally allocate the resources, e.g., changing illumination or routing depending on the distance to the user.

Spatial structure is particularly susceptible to methods based on grids, spatial convolutions, or graph representations when the structures are non-Euclidean or dynamic. Graph neural networks (GNNs) and spatiotemporal convolutional networks have demonstrated themselves to be useful in such modeling in the use of human-centric analytics such as crowds, indoor navigation, and collaborative robotics. Notably, spatial relations do not persist unchanged over time, interactions change, and thus spatial reasoning cannot be solved without time modeling in a given environment to realize prediction and dynamic behavior in CPHSs.

6.2.3 Multimodal Data Fusion (Audio, Visual, Sensor, Text)

CPHSs are multimodal in nature and depend on distinct streams of data, each of which captures some particular characteristics of the state of the system and the human user. Visual data give people depth perception, object identification, and figurative gesture comprehension; audio finds verbal communication, emotional response, and surrounding context; sensor information gives reports on motion, heat, or body state; and text or command input reflects goals or inclinations. It is very important to have a fusion of such modalities to construct a more complete picture of context and increase the accuracy, responsiveness, and robustness of a given system, particularly in environments that are rather unpredictable.

The issues in the integration of multimodal data include synchronization, noise tolerance, and temporal resolution. Fusion strategies can include early fusion (raw feature fusion), late fusion (combining the output of separate models), and hybrid fusion (overlapping representations). The modern approaches to align and correlate modalities are cross-modal attention, multistream networks, and the encoder–decoder framework. Multimodal fusion allows the systems to become more adaptable and intelligent, as it allows the systems to analyze complex human behavior in real time, providing more consistent and natural interaction when carried out properly.

6.3 Deep Neural Network Architectures for CPHS

Spatiotemporal CPHSs are sensitive to the development of high dimensional data that change over time and space and are largely dependent on progressive deep learning architectures to analyze and make meaning of the same. The architecture of the used neural networks greatly affects the property of the system to learn new and complex patterns, adapt to human behavior, and arrive at reliable decisions [4]. Different architectures will achieve some strengths in temporal modeling, spatial representation, and multimodal fusion, depending on the nature of data and requirements of a specific application. In this section, specific concepts under deep learning commonly deployed in CPHSs will be investigated, and how they act to support intelligent, responsive, and human-conscious functionalities will be explored.

6.3.1 Temporal Modeling with RNNs, LSTMs, and GRUs

The crucial importance of finding the sequence and timing of the events is essential not only to be able to predict and react to those events, but also to temporal modeling, which is very central in the use of CPHSs. The basic structure behind such tasks is referred to as the RNN, which has architectures that keep the memory of the previous inputs *via* the recurrent connections [5]. Nonetheless, the standard RNNs have problems with long-term dependency, particularly because of the vanishing gradients problem, and thus they are not adequate in modeling long-range human behavior or physiological processes as time elapses.

LSTM networks and GRUs were developed to overcome these drawbacks. This is because they have built in the gating processes through which the network can choose to remember or even exercise the ability to forget information, making them more efficient when learning long-term cases. LSTMs and GRUs have had application success in CPHS areas such as activity recognition, real-time monitoring, and anomaly detection [6]. Their capability to represent the nonlinear, context-rich time tendencies makes them invaluable to the systems required to predict or adjust to the developing human habits [7].

6.3.2 Spatiotemporal Fusion Using 3D-CNNs and ConvLSTMs

Spatiotemporal fusion encompasses combining both of the characteristics, spatial and time, to achieve a more comprehensive feeling of dynamic settings. 3D convolutional neural networks (3D-CNNs) are an extension of 2D convolution into the time dimension, i.e., a process of using 3D-convolution-based methods to process video or volumetric data [8]. Such models are competent in representing the short-term motions and spatiotemporal correlations such as gesture perception or recognition in intelligent surveillance applications. In cases that demand a long-term relationship and also a spatial reasoning, ConvLSTMs are better adapted. They combine both the convolutional operation and the LSTM cell, where the model can maintain spatial information when learning the temporal information. This renders ConvLSTMs particularly suitable to such uses as motion prediction of humans, medical surveillance, and interactions, where motion and context are both paramount. Such architectures provide a tradeoff between accuracy and explanation in elaborate CPHS settings.

6.3.3 Attention Mechanisms and Transformer-Based Models

Attention mechanisms transformed the field of deep learning by enabling the models to zero in on the most relevant sections of the input sequence and enabling better accuracy as well as interpretability. As opposed to the traditional recurrent models, attention allows input features to be weighed dynamically and processed in parallel, a feature that is specifically valuable in CPHSs because of the asynchronous occurrence of events and the irregular nature of human behavior. Self-attention modules can select useful segments of time or regions of space, making the model more context-prioritized.

Transformer models, which have been proposed in natural language processing, have gained extended use on spatiotemporal problems. They are designed using attention mechanisms exclusively and, hence, can model nonrecurrent dependent relationships. They have been effective in activity recognition, emotions detection based on speech, and even multimodality data fusion *via* transformers. Their high degree of scalability, their strength against the length of the sequence, and a proliferating family of variants (e.g., vision transformers, TimeSformers) make them an excellent solution to the CPHS applications where fine-grained temporal and spatial reasoning is a critical component.

6.3.4 Hybrid Models for Complex Human-Centric Tasks

There is no common architecture available to tackle all aspects of CPHSs mixing with different data modalities and tasks extending across a wide range of time and space scales. Hybrid models can be strengthened by combining various neural architectures, e.g., CNNs are used to extract spatial features and LSTMs or transformers are used to model temporal structures. These combinations enable systems to analyze visual information and at the same time have an idea about the changing state of human behaviors or environmental changes [9].

Hybrid models find their specific utility in the cases of multimodal emotion recognition, smart healthcare diagnostics, and intelligent transportation systems where the human behavior context is a product of the physical environment and changing time. A combination of models in these architectures allows the flexibility of the systems, a better inference of predictions, and resistance to any noise and missing data. They are the next generation of developing strong, adaptive, and context-aware CPHSs able to enable real-time, adaptive, and human-friendly interaction.

6.4 Architectures for Spatiotemporal Modeling

CPHSs need special neural network architectures to model spatiotemporal data, taking care of where and when things happen. Such architectures have to deal with temporal water flows, spatial geometries and relationships, movement, change, volatility, and dependencies between system objects and humans [10]. CPHSs grow increasingly sophisticated and data-intensive, and it is therefore vital to derive significant patterns out of structured, multimodal, and time-sensitive inputs. A brief discussion of major deep learning architectures—each of which is different along the degrees to which they excel in spatiotemporal modeling—is done in this part.

6.4.1 Recurrent Neural Networks (RNN, LSTM, GRU)

One of the first models of deep learning applicable to sequential data are RNNs. They only perform one step forward on input data and have an internal state to reflect the temporal dependency. Nonetheless, when learning long-term dependence, RNNs sometimes have problems such as the vanishing gradient, thus they are not easily applicable to modeling human behavior patterns (long or complex type) in CPHSs. In an attempt to address these shortcomings, more sophisticated versions (LSTM networks and GRUs) were found. Such architectures possess gating mechanisms that allow the storing of significant characteristics over extended sequences. LSTMs and GRUs are very popular in their application in CPHSs, with gesture recognition, human activity forecasting, and real-time monitoring as the prime examples where the time and order of actions are crucial to understand. Their capability of modeling dynamic user behaviors has made them highly useful in human-in-the-loop systems.

6.4.2 Convolutional Neural Networks for Spatial Patterns

CNNs are the architecture of choice when it comes to modeling the spatial relationship of (structured) data, and especially visual data, e.g., image or video frames. CNNs use convolutional filters to extract local features such as edges, textures, and features of objects, and as a result, they can effectively learn spatial hierarchies [11]. CPHS tasks where CNNs have been utilized include facial recognition, environment mapping, and posture detection, in which the physical layout of features is very important. Although usually restricted to the modeling of static imagery, CNNs can be further modified to the spatiotemporal domain, e.g., through processing of the sequences of frames or 3D convolutions. CNNs offer an excellent architecture when paired with models in other dimensions such as temporal data, whether in an LSTM structure or in an attention model. This renders them a useful component in a system where decisions are made based on visual-spatial signals, i.e., intelligent surveillance or assistive robotics.

6.4.3 Graph Neural Networks for Structured CPHS

Structured data are present in many CPHS environments and are not easily expressible as a grid structure, e.g., the topology of a sensor network, human interaction, or transportation systems. GNNs are intended to perform tasks on these irregular structures by modeling the data as nodes and edges, identifying complex relational features [12]. They make CPHS dynamic development and understand the local or global dependencies to have the capacity to reason in the context-aware environment where entities are not in relation to each other in a trivial way. Multiagent coordination, activity recognition in crowds, and smart building management are some of the unique applications of GNNs in CPHSs. As one example, a GNN can be used to learn how people and devices in a smart home interact to determine the intention or maintain optimal resource utilization. GNNs are well suited to structured, distributed systems in which the topology itself is both dynamic and significant.

6.4.4 Transformer-Based Models for Spatiotemporal Learning

Transformer models have drastically changed the world of deep learning, because they model complex dependency in a way where self-attention is applied. In contrast to RNNs, data can be processed in parallel in transformers, and certain input sections are more highly concentrated on by the model dynamically compared to others—transformers are extremely efficient when run on long sequences [13]. This is particularly beneficial in CPHSs where streams of data may be only *ad hoc*, chaotic, and subject to a wide array of variables.

Spatiotemporal transformers expand this by adding in spatial and temporal attention layers that enable them to learn complex patterns in their representations in space and time at once. These models have proven to be very powerful in activity recognition, speech and gesture recognition, and multimodal fusion. Their flexibility, generality, and scalability make transformers one of the dominant architectures of the next generation CPHS systems that may need to reason about complex evolving human conditions [14].

6.5 Training Deep Networks for CPHS

The process of training DNNs in the context of CPHSs is of a crucial yet complicated nature that has a lot of challenges peculiar to spatiotemporal and human-centered applications [15]. These systems are in contact with dynamic environments and unpredictable human behavior, and they strongly depend on sound data-driven learning mechanisms. Proper training is also not just about using the proper model architectures but also about the proper design of datasets, suitable learning paradigms, and adaptation strategies so that the model can generalize to different contexts.

This section explores the core modules of the training pipeline: how to collect and annotate data; what kinds of learning frameworks by which to train deep models; transfer learning and domain adaptation methods; and how an existing model can be modified to fit new CPHS settings [16].

6.5.1 Dataset Design and Annotation Strategies

Quality data are the pillar of any deep learning model. In CPHSs, the purpose of designing appropriate datasets is to record intricate spatiotemporal dynamics that are frequently multimodal—in terms of, say, visual, auditory, physiological, and sensor-level signals. With all this variety of inputs, data collection is not only time-consuming but technical as well. Systems are required to log transactions in real time or in varied conditions and in synchronous perpetual information collected in numerous sources.

The annotation process is also critical, and it provides semantic meaning to the raw data. In CPHSs, these activities can be labeled as emotion states, intent tags, physiological conditions, or spatial positioning. Annotation is prone to error and may be time-consuming, especially in ambiguous or overlapping human behaviors. To overcome that, active learning methodologies, crowdsourcing, and semiautomated annotation tools become more common to label data fast and retain quality at the required level.

6.5.2 Supervised, Unsupervised, and Self-Supervised Training

Training paradigms of CPHSs may be diverse, depending on the access to labeled data and application types. The most widespread use is supervised learning, in which models are learned based on input–output pairs obtained *via* annotated datasets. Such a paradigm works well where there is a large amount of data that are labeled but lacks in conditions that need personalization or are real-time such that new data may not be labeled conveniently. To address these limitations, unsupervised learning methods are used to infer latent patterns and structures in unlabeled data. The approaches are more applicable in exploratory assignments, anomaly learning, and the grouping of human actions without prior category formulations. These could be autoencoders, generative adversarial networks, and contrastive learning methods [17].

One prospective development of the past few years is self-supervised learning, a technology where the space between supervised learning and unsupervised algorithms is filled. In such a modification, the data are used to create pseudolabels or prediction tasks for the system, such as predicting the subsequent frames of a video or masked sensor inputs. Self-supervised methods are becoming popular in CPHSs due to the aspect that they allow pretraining on large unlabeled datasets, which can subsequently be fine-tuned in a certain task with little labeled data. This not only decreases annotation costs but improves generalization to users and settings.

In addition, hybrid learning arrangements, including semisupervised learning and multitask learning, can utilize small quantities of labeled information and enormous numbers of unlabeled corpora. Such approaches are especially useful in CPHS applications such as healthcare surveillance and assistive robotics, as well as smart environments, where keeping the data labeled consistently is a high-cost and mission-critical task [18].

6.5.3 Transfer Learning and Domain Adaptation

CPHS applications tend to differ widely in physical settings, the populations being served, hardware infrastructure, and interaction patterns. Consequently, agents learned in one environment (e.g., a smart hospital) can be unlikely to generalize to others (e.g., a home-based healthcare system) that experience a distributional shift. Transfer learning is a viable way out because it involves reusing already learned models and reusing the knowledge for a new task or new environment with minimum retraining. In transfer learning, a model is trained on a large generic dataset (e.g., ImageNet or some publicly available dataset of human activity) and then fine-tuned on a smaller dataset that is specific to a task. This method makes a major improvement in terms of computing costs and training time, and performance in low-resource settings. In CPHSs, transfer learning is pervasive in terms of personalizing systems to individualized users, adapting to new sensors, or applying to new tasks.

Simple transfer learning might not, however, be adequate where the source and target domains are not very similar. Here, domain adaptation schemes are applied that reduce the difference between domains.

They seek to realign the distributions of features in the source and target data (typically *via* adversarial training, domain-invariant representations, or fine-tuning with a regularization penalty). In a scenario, consider that domain adaptation enables a gesture recognition system to be used in a real environment with little consequence to its performance, which might have been trained in the lab. Additionally, online domain adaptation is also effective in CPHSs in particular, where there is a continual learning and adaptation of new, unlabeled data when deployed. The ability makes the models robust and precise even in a nonstationary environment, conducive to long-term autonomy and adaptability. With CPHS systems on their path to increased personalization and real-time decision-making, transfer learning or domain adaptation will play an invaluable role in allowing scalable and sustainable deployment.

6.5.4 Optimization Techniques and Model Tuning

Deep learning models in the case of CPHSs do not work based on the architecture alone or on training data alone; they work based on whether the model is optimized and fine-tuned. Because the CPHS is real-time, context-aware, and human-centric, the optimization methods should balance accuracy, generalization, latency, and robustness [19]. Such issues as varying length of inputs, noisy input data, and data scarcity make the thoughtful construction of models important in getting quality results in spatiotemporal systems. Here, this section dwells on such essential aspects of the optimization process as the choice of loss functions specific to spatiotemporal data, the means of preventing overfitting, and the key principles of optimization hyperparameters to increase the effectiveness and flexibility of a model working in changing conditions.

6.5.5 Loss Functions for Spatiotemporal Tasks

Loss functions are the objective alternatives that allow learning within a model. The application of CPHSs in the spatiotemporal domain requires a profound loss selection because the adoption of a suitable choice of the loss can determine the capacity of the model to represent time continuity, spatial connections, and human-related variability. Common loss functions such as mean squared error, cross-entropy, and categorical hinge have been found to be inadequate for spatiotemporal modeling because these common functions do not necessarily capture the peculiarities of spatiotemporal modeling.

In the temporal tasks of sequence prediction or time-series forecasting, loss functions need to consider temporal coherence. As an example, dynamic time warping and time smoothness loss can be used to punish discontinuities in time-dependent outputs. Loss functions better able to assess the accuracy of predictions in a spatial setting include, in spatial realms, intersection over union or heatmap-based losses in instance, gesture, or motion recognition. Focal loss and weighted cross-entropy can aid in distribution in classification tasks where the labels are unbalanced, particularly when it involves rare human behaviors. The custom loss functions applied by multimodal CPHS applications might encompass the combination of multiple objectives (e.g., accuracy, temporal consistency, and localization). For human-in-the-loop systems, a key to making effective use of hybrid loss functions is designing these functions to exhibit the same priorities as in the real world, including responsiveness, safety, and personalization.

6.5.6 Regularization and Overfitting Control

Variability and the scarcity of labeled data are persistent problems in CPHSs because overfitting is frequent. Good models on training sets tend to be bad on new users, devices, or conditions. It is vital to counter such a problem using regularization approaches to enhance the robustness of models. L1 and L2 regularization, which punish excessively complex weight schemes to facilitate the simplification of model behavior, are the most popular among them. Dropout is a stochastic regularization method under which, during training, a random subset of neurons is removed, which makes the network learn more generalized features. In spatiotemporal networks, dropout may be generalized to the temporal dimensions (e.g., variational dropout) to avoid dependence upon particular time points. In a parallel way, data augmentation, accomplished e.g., by time warping, jittering, flipping, and injecting noise, simulates intrauser and intrasetting variability and enhances cross-user and intersetting generalization.

Besides these, early stopping using a validation loss is also a feasible way to address overfitting, especially in edge systems or a mobile device constrained by limited computation capabilities. Other more sophisticated techniques such as Bayesian regularization, label smoothing, or adversarial training become popular in safety-critical applications of CPHSs as well. But such methods are not only a means of increasing the performance of the model but also make it more resilient against unexpected input and data drift.

6.5.7 Hyperparameter Optimization Methods

In deep learning, hyperparameter tuning is sometimes viewed as both an art and a science, and in CPHSs, it is exponentially increased by the variation of system settings and user actions. Other hyperparameters that can greatly influence accuracy and computational efficiency are the learning rate, size of a batch, number of layers, magnitude of kernels, and length of temporal windows. Manual tuning is laborious and likely to be less than optimum, as in the more demanding spatiotemporal models [20]. Random search and grid search are conventional but straightforward methods in terms of searching hyperparameter space. Although grid search attempts all predetermined combinations in a completely exhaustive manner, random search discovers good results more quickly by taking samples throughout the space. More efficient optimization methods such as Bayesian optimization and tree-structured Parzen estimators leverage probabilistic models to learn what combinations of hyperparameters will likely perform better, further limiting the number of trials to perform.

Automated hyperparameter tuning, evolutionary algorithms, and optimizers based on reinforcement learning are also coming into use as highly effective automated hyperparameter tuning approaches, especially in AutoML frameworks. Such methods can adjust the search space online, using real-time performance measurements, which makes them suitable for the CPHS systems, which are faced with changing dynamics.

6.6 Edge and Real-Time Deployment Considerations

Because more and more CPHSs start shifting out of centralized cloud infrastructure and into distributed edge computing landscapes, there is a high demand in the capability to operate deep learning models in real time and resource-strained contexts. Edge computing facilitates on-device decision-making with low latency and minimizes the need to connect to the cloud, and it is an indispensable feature of remote health monitoring, smart homes, autonomous cars, as well as wearable devices [21]. Nevertheless, complex models present distinctive complications for deployment on edge devices in regard to computational effectiveness, energy usage, memory utilization, and inference latency duration. In this part, the main concerns of edge deployment within CPHSs are handled with a concentration on three main points: methods of compression of models, training on edge devices, and latency-conscious inference in systems where people are included as the constituents of the decision loop.

6.6.1 Model Compression and Pruning

The common approach of deep learning networks is the use of computationally intensive solutions, favoring the use of normal computers and systems with abundant resources [22]. Model compression methods are used to make models smaller and less complex, as well as to consume less memory without a significant loss of accuracy. Popular ones are weight pruning, quantization, low-rank matrix factorization, and knowledge distillation. The methods facilitate the possible application of real-time inference on an embedded platform, e.g., a microcontroller, smartphone, or IoT device [23].

Pruning is used to establish and delete unnecessary or nonessential redundant weights or neurons, creating a lighter and sparser model. These are performed in either a static way after training or dynamically during training by means of structured or unstructured pruning techniques. Quantization can also shrink models because representations of weights and activations are encoded in lower-precision formats (e.g., INT8 as an alternative to FP32) and enhances performance, truncating the memory bandwidth demands. In the meantime, knowledge distillation is training a smaller (student) model to imitate the predictions of a larger and more accurate (teacher) model—sometimes being able to reach near-optimal accuracy with a few hundred times fewer parameters. In the case of the compression of CPHS applications, trade-off analysis is an important element. Although minimizing model size can increase deployment efficiency, this minimization should not impede a model from making safety-critical or context-dependent decisions in areas where human interaction is in real time.

6.6.2 Training on Edge Devices

Most deep learning models are trained on high-performance computing infrastructure; however, there has been a recent interest in making it possible to also train on-device, especially in applications where personalization or continuing adjustments are of importance. In CPHS applications, user behaviors and settings change with time; it is possible to learn these behaviors or settings by training at the edge devices themselves, instead of requiring the sending of sensitive information to the cloud. This aids privacy, responsiveness, and optimization to the user [24].

Lightweight learning algorithms, model structures, and architectures enable edge-based training in low-resource settings. Such techniques will be applicable in these contexts and include few-shot, incremental, and online learning. Moreover, federated learning has become one of the central paradigms, where a large number of edge devices cooperatively learn and optimize a global model and do not exchange raw data. This will be especially useful in the cases of CPHSs used with personal or medical data, because it will comply with data protection regulations such as the General Data Protection Regulation (GDPR) or the Health Insurance Portability and Accountability Act (HIPAA). Although this has a number of benefits, there are limitations to training on edge devices. The same issue of limited computing resources, battery power, and memory capacity demands the optimization of the model choice, data batching, and learning rate control. Also, on-chip training should be handled with caution to not get completely out of control, overheat, cause discomfort to the user, or crash the system, particularly in wearable or implanted units.

6.6.3 Latency-Aware Inference for Human-in-the-Loop Systems

Latency in CPHSs does not only impact the performance measurements but also usability, safety, and user confidence directly. Latency-aware inference is a method of designing so that the system can input and produce meaningful output in a time frame that enables real-time decision-making and interpersonal interaction. As an example, any health monitor alert system or a robot controlled by gestures can be dysfunctional and even unsafe in the cases where its response time exceeds the control of a few hundred milliseconds. There are a number of approaches that can be taken to attain low-latency performance. These are using lightweight architectures (e.g., MobileNet, SqueezeNet), low input resolution, low frame rate configurations, and asynchronous models of execution. In addition, some models can undergo partition and inference along pipelines that enable partial model execution on diverse processors or at the edge-cloud continuum with a balance between speed and precision. The throughput can further be enhanced on-device through hardware acceleration with graphics processing units, tensor processing units (TPUs), or specialized AI chips (such as the Google Edge TPU or NVIDIA Jetson).

In the case of human-in-the-loop systems, it is vital that latency be predictable and bounded [25]. These applications, in contrast with a strictly automatic system, are required to match the human time scale of thinking and reaction. The goal of real-time feedback loops must be to maximize both speed and other properties, such as clarity, reliability, and adaptability. The problem is even more crucial in such applications as collaborative robotics, assistive technologies, and augmented reality, where there is a close interconnection between the machine response and the human activity.

6.7 Future Trends and Research Opportunities

Due to the potential evolution of CPHSs to be more intelligent, more autonomous, and more integrated with human life, a wide variety of new directions of research as well as technological challenges are arising. CPHSs are evolving into a more flexible, ubiquitous, and data-driven environment, but this added complexity poses some serious issues about ethics, trust, and actual practicality. As the field continues to rely on spatiotemporal sensing and deep learning, it will be important in the future to begin to focus on not only performance but also responsibility, transparency, and scalability. This section talks of the emerging research areas and open problems in the CPHS area, in particular, the importance of ethical and privacy concerns, the necessity of robust and interpretable models, and the necessity of scaling these solutions to real-world use within diverse and dynamic settings.

6.7.1 Ethical and Privacy Concerns

The incorporation of humans into the computational and actual loop brings heavy ethical challenges in CPHS design and implementation. Such systems are increasingly based on the persistent gathering and analysis of personal and behavioral data, such as location, gestures, speech, facial expressions, and physiological signals, which give way to much more serious issues of privacy, consent, and data ownership. Making sure that one is informed and can manage the usage of his or her own data is both a requirement of ethical standards and a legal one of regulations such as the GDPR, HIPAA, and even the Digital Personal Data Protection Act in India.

CPHS research must take privacy-preserving methods into account in the future; these will include differential privacy, homomorphic encryption, and federated learning, each of which allows computations to be made without revealing sensitive information about users. Besides, privacy-by-design frameworks such as data minimization, anonymization, and transparency will also play an important role in developing the trust between the population and companies. Ethical design is also the process of considering all possible biases in the data or models that may yield discriminatory or exclusionary outcomes in sensitive systems and applications such as healthcare, recruitment, or public safety. Ethical AI in CPHSs should also expand to algorithmic responsibility, fair access to new technology, and making sure that the technologies complement human beings and do not substitute for them. As technology makes itself more autonomous, the definition of where machine and human decision-making begin and

and will be a central territory of philosophical, social, and technological discussions.

6.7.2 Robustness, Explainability, and Trustworthiness

CPHSs could not remain opaque or have their decisions trusted in sensitive human-centric areas without being able to show trustworthy, explainable behavior and demonstrated robustness. Real-life circumstances differ from closed environments because they are not predictable, nonstationary, or noisy. Models trained using a given assumption might not fare well when they are used in varied physical conditions, cultures, as well as populations of users. To combat this issue, it is important to come up with models that can withstand the out-of-distribution inputs, missing values, sensor errors, and even the adversarial attacks. Such techniques as ensemble learning, input normalization, adversarial training, and uncertainty modeling may be used to improve robustness. Technical robustness is, however, not enough because a user should know the rationale behind a particular decision that a system makes. This has translated into the increasing popularity of explainable AI—explainable AI approaches—that envisage producing model predictions that are explainable to the nontechnical stakeholder as well as the technical stakeholder. For example, saliency maps, attention visualizations, and surrogate models (such as LIME [local interpretable model-agnostic explanations] and SHAP [SHapley additive explanations]) can give human-interpretable understanding of model behavior.

Finally, accuracy is not the only factor that builds trustworthiness in CPHSs based on transparency, consistency, fairness, and responsiveness. It is very crucial, especially in such applications as autonomous healthcare, collaborative robotics, or emotion-sensitive applications where humans use the system as a guide or in making their decisions. The user should not give the system too much or too little trust, which is a considered research field at the intersection of human psychology, interface design, and AI.

6.7.3 Scalability for Real-World Deployments

Despite having achieved considerable progress in laboratory and prototype environments, some of the CPHS solutions grapple with the challenges of scaling to practical applications. Scalability is over and above computation performance; it includes flexibility to be scaled to accommodate multiple users, devices, environments, and application spaces. Scalability of CPHSs can only be realized through careful system design to support modularity and interoperability as well as continuous learning. As an example, a smart health-monitoring system should operate in a hospital, home, and wearable as well as adjust to patient-specific behavior patterns and requirements.

The wide deployment should support cloud-edge hybrid systems, scalable training pipelines (e.g., through distributed learning or federated updates), and efficient communication protocols. Moreover, platform-independent development and APIs and data standardization will play a vital role in making sure that the elements of CPHSs can be reused and integrated with each other across verticals, such as healthcare, transportation, and education.

The opportunity is also high with regard to the development of self-adaptive systems that would be able to oversee their own performance and commit the necessary updates or retraining. The synergy of online learning, the feedback loop, and human supervision of systems encountered through monitoring can make the CPHSs more resilient in matters concerning scale-related degradation. Moreover, making sure that CPHS technologies will be applicable to real-life, everyday practice will necessitate designing CPHS devices considering their energy efficiency, low cost, and regulatory standards, especially in low-resource areas or emerging ones.

6.8 Conclusion

This chapter discussed the nature of the relationship between DNNs and CPHSs in spatiotemporal complexity and human behavior management. Starting with CPHS architecture and human-in-the-loop feedback, it focused on the necessity of intelligent and adaptive systems. RNNs, LSTMs, 3D-CNNs, GNNs, and transformers are different models that have been mentioned to process multimodal data. Some approaches to training were presented, such as supervised, self-supervised, domain adaptation, and federated learning in response to data scarcity and personalization. Loss functions, regularization, and hyperparameter tuning aspects of optimization have been analyzed. Such deployment aspects as edge computing, model compression, and latency-aware inference were taken into account. Lastly, the chapter touched on the future obstacles in ethics, privacy, explainability, and scalability, thus making CPHSs a revolutionary entity in fields such as healthcare, smart cities, and transportation.

AI disclosure statement: The author confirms that no artificial intelligence (AI)-generated content or AI-assisted writing tools were used in the preparation of this chapter (including summaries and the abstract). All text, analysis, and conclusions are the result of the author's original work.

References

1. Musa, A.A., Hussaini, A., Liao, W., Liang, F., Yu, W., Deep Neural Networks for Spatial-Temporal Cyber-Physical Systems: A Survey. *Future Internet*, 15, 199, 2023.
2. Müller, H.A., The Rise of Intelligent Cyber-Physical Systems. *Computer (IEEE)*, 50, 7, 2017.
3. Shao, Z., Liu, J., Ding, Z., Chen, M., Jiang, N., Spatio-temporal Properties Analysis for Cyber-physical Systems. *18th Int. Conf. on Engineering of Complex Computer Systems*, Singapore, pp. 101–110, 2013.
4. Deshmukh, J.V., Kapinski, J.P., Yamaguchi, T., Prokhorov, D., Learning Deep Neural Network Controllers for Dynamical Systems with Safety Guarantees. *IEEE/ACM Int. Conf. on Computer-Aided Design (ICCAD)*, Westminster, USA, pp. 1–7, 2019.
5. Mienye, I.D., Swart, T.G., Obaido, G., Recurrent Neural Networks: A Comprehensive Review of Architectures, Variants, and Applications. *Information*, 15, 517, 2024.
6. Wu, X., Liu, Z., Wang, L., Spatio-temporal degradation model with graph neural network and structured state space model for remaining useful life prediction. *Reliab. Eng. Syst. Saf.*, 256, 110770, 2025.
7. Almagrabi, A.O., A Deep CNN-LSTM-Based Feature Extraction for Cyber-Physical System Monitoring. *Comput. Mater. Continua*, 76, 2079, 2023.
8. Ren, Z., Xiao, X., Nie, H., Empowering Efficient Spatio-Temporal Learning with a 3D CNN for Pose-Based Action Recognition. *Sensors*, 24, 7682, 2024.
9. Graessler, I. and Poehler, A., Human-centric design of cyber-physical production systems. *Procedia CIRP*, 84, 251, 2019.
10. Kumar, R., Bhanu, M., Mendes-Moreira, J., Chandra, J., Spatio-Temporal Predictive Modeling Techniques for Different Domains: A Survey. *ACM Comput. Surv.*, 57, 42, 2025.
11. Wang, Q., Parker, P.A., Lund, R., Spatial deep convolutional neural networks. *Spatial Stat.*, 66, 100883, 2025.
12. Khemani, B., Patil, S., Kotecha, K., *et al.*, A review of graph neural networks: Concepts, architectures, techniques, challenges, datasets, applications, and future directions. *J. Big Data*, 11, 18, 2024.
13. Yeung, C., Sit, T., Fujii, K., Transformer-based neural marked spatio-temporal point process model for analyzing football match events. *Appl. Intell.*, 55, 335, 2025.
14. Mao, Y., Zhang, G., Ye, C., A Spatio-temporal Graph Transformer driven model for recognizing fine-grained data human activity. *Alexandria Eng. J.*, 104, 31, 2024.
15. Ma, Z., Mei, G., Piccialli, F., Deep Learning for Secure Communication in Cyber-Physical Systems. *IEEE Internet Things Mag.*, 5, 63, 2022.
16. Abosuliman, S.S., Deep learning techniques for securing cyber-physical systems in supply chain 4.0. *Comput. Electr. Eng.*, 107, 108637, 2023.

17. Cao, Y., Yan, J., Feng, H., Wang, T., Qiao, Y., Peng, T., SCAAE: Using Self-Supervised Contrastive Learning in Adversarial AutoEncoder for Anomaly Detection of Multivariate Time Series in Cyber Physics Systems. *China Autom. Congr. (CAC)*, China, pp. 8102–8107, 2023.
18. Wickramasinghe, C.S., Amarasinghe, K., Marino, D.L., Rieger, C., Manic, M., Explainable Unsupervised Machine Learning for Cyber-Physical Systems. *IEEE Access*, 9, 131824, 2021.
19. Sheikh, Z.A. and Singh, Y., OLM-CPS: Optimized Learning Models for the Security of Constrained Cyber Physical Systems. *Procedia Comput. Sci.*, 259, 1845, 2025.
20. Pravin, P.S., Tan, J.Z.M., Yap, K.S., Wu, Z., Hyperparameter optimization strategies for machine learning-based stochastic energy efficient scheduling in cyber-physical production systems. *Digit. Chem. Eng.*, 4, 100047, 2022.
21. Sánchez, J.M.G., Jörgensen, N., Törngren, M., Inam, R., Berezovskyi, A., Feng, L., Fersman, E., Ramli, M.R., Tan, K., Edge Computing for Cyber-physical Systems: A Systematic Mapping Study Emphasizing Trustworthiness. *ACM Trans. Cyber-Phys. Syst.*, 6, 28, 2022.
22. Li, Z., Li, H., Meng, L., Model Compression for Deep Neural Networks: A Survey. *Computers*, 12, 60, 2023.
23. Nasif, A., Othman, Z.A., Sani, N.S., The Deep Learning Solutions on Lossless Compression Methods for Alleviating Data Load on IoT Nodes in Smart Cities. *Sensors*, 21, 4223, 2021.
24. Archana, R. and Jeevaraj, P.S.E., Deep learning models for digital image processing: a review. *Artif. Intell. Rev.*, 57, 11, 2024.
25. Wu, X., Xiao, L., Sun, Y., Zhang, J., Ma, T., He, L., A survey of human-in-the-loop for machine learning. *Future Gener. Comput. Syst.*, 135, 364, 2022.

7

Digital Twin–Driven Model Architecture for Cyberphysical Systems Toward the Development of Testbed

Asha Kumari Ablikatte[1]*, Prajwal Balakrishna[1], Saravana Kumar E.[1] and P. Bindu Madhavi[2]

[1]Department of Computer Science and Engineering, The Oxford College of Engineering, Bengaluru, Karnataka, India
[2]Department of Computer Science and Engineering, T. John Institute of Technology, Bengaluru, Karnataka, India

Abstract

A digital twin (DT) mirrors real-world systems and creates a virtual replica of devices, applications, or even humans. It enables real-time simulation, supervision, and fine-tuning by maintaining a continuous exchange of data among the tangible and digital environments. In cyberphysical systems (CPSs), DTs are used extensively as mirrors to monitor, analyze, and control the real systems. This chapter provides a DT-driven, five-layered architecture for CPSs. The DT architecture aims to provide a safe testing environment, predictive analysis, and risk-free validation. The five-layer DT architecture consists of physical, cyber, DT, synchronization, and experimental layers. This architecture can be applied across various domains, including manufacturing, healthcare, and smart city development. The testbed includes real-time communication protocols such as message queuing telemetry transport and open platform communications unified architecture for providing reliable data transfer, monitoring, and synchronization. The experimentation validation is reserved as future work; theoretical evaluation suggests that the architecture reduces the developmental cost and testing cost of CPSs, and minimizes the risk to physical components. This chapter lays the groundwork for developing hybrid digital testbeds for cyberphysical human systems and their efficient management.

**Corresponding author*: asha.hebbar@gmail.com; ORCID: 0000-0002-7556-5933

Anitha Velu, Prasanth Aruchamy, Raghu Ramamoorthy, Rajesh Kumar Dhanaraj and Manjula Arunraj (eds.) Cyber Physical Human Systems: Emerging Trends and Applications, (153–178) © 2026 Scrivener Publishing LLC

Keywords: Digital twin technology, cyberphysical systems, digital twin testbed, five-layered DT architecture, Industry 5.0

7.1 Introduction

Recent technological innovations have accelerated the transition from Industry 4.0 to Industry 5.0, reflecting a paradigm shift from conventional cyberphysical systems (CPSs) to human-centric cyberphysical frameworks [1, 2]. The primary aim of Industry 5.0 is to promote seamless collaboration between humans and intelligent machines, which is embodied in collaborative physical human–cyber systems (CPHCSs). This collaboration enables industries to increase the overall productivity, accountability, and also the adaptability of CPHCSs in various industry-based applications. Some key applications of cyberphysical human systems (CPHSs) and CPSs are in several key domains, including but not limited to manufacturing, construction, medicine, energy, automotive, smart cities, and farming.

In the digital era, managing real-world systems can be changed drastically by digital twin (DT) technology. It shifts the paradigm of how we process, manage, and interact with the physical systems by creating a virtual copy by providing integration with physical and virtual systems. This technology progressed from theoretical concepts to practical application, and various sectors have adopted it widely [3]. The concept of DT is based on developing a digital partner for physical components, using state-of-the-art modeling techniques to display the real-time status of physical components. The DTs cannot be distinguished from their physical counterparts [4]. From the micro atomic level to the macro level, the DT represents its physical counterpart. The information that can be obtained from observing the real-world object can also be obtained from observing the DT. Digital twins range from component twins, representing a single component of a complex system, to asset twins through representing the entire system [5]. The DT technology can be applied in various sectors, such as manufacturing, construction, energy, automotive, healthcare, urban planning, etc.

7.1.1 Industry 5.0

Industry 5.0 represents the next stage in the evolution of Industry 4.0, building upon its foundational principles and technologies while introducing a significant transformation in the way industry functions and engages with technology. Industry 4.0 tailored CPSs, cloud computing, and supports complete automation from industry to production. This was introduced by Germany in the year 2011. The key problem in Industry 4.0 was that humans were out of the loop, as it focused on complete automation and the elimination of humans from the industrial systems. To overcome the above-stated problem, Industry 5.0 was introduced into automated systems. Industry 5.0 is an alliance between humans and the machines, hence building an intelligent system incorporating both by ensuring that humans are in the loop. Table 7.1 compares the key differences between the above-stated industries.

Table 7.1 Comparative overview of Industry 4.0 and Industry 5.0.

Areas	Industry 4.0	Industry 5.0
Key goal	Automation and efficiency	Human–artificial intelligence (AI) collaboration
Human role in the industry	Minimal, humans are replaced in most scenarios	Human-centric and decisions are made by humans
Technology use	Cyberphysical systems, cloud computing, Internet of Things (IoT), smart factories	Edge computing, federated learning, digital twins, collaborative robots (CoBot), explainable AI (XAI)
Decision-making	Automated and fully machine-driven, hence reduced human control	Human-centric, hence important decisions and the workflows are controlled by the humans in the loop
Scalability	Global	Local and decentralized, hence useful in real-time systems, thereby making it less scalable compared to Industry 4.0
Data processing	Centralized	Decentralized
Ethical AI consideration	Limited to very less ethics in usage of AI and data for training AI	Strong emphasis on the set of ethics and rules to be followed governing AI
Sustainability	Secondary concern	Core objective of Industry 5.0

Some of the main characteristics that distinguish Industry 5.0 from Industry 4.0 are: human-centricity, where the primary goal is to develop intelligent systems and machines that keep humans at the core of the design process. These technologies emphasize collaboration between people and mechanical systems, confirming that automation enhances rather than replaces human involvement. This means that humans have control over the entire process and hence ensures that the industry environment is more trustworthy and more efficient as compared to Industry 4.0. This also helps to enable accountability to the industry and hence ensure that the quality of the product is maintained in the industrial settings.

Similarly, resilience enables industries to build a set of robust systems that are adaptive in nature, which means that they can adapt to any scenario and also to any kind of disruptions such as pandemics, political tensions, or shocks in the supply chains. This ensures that the industries are more resilient to the external world and hence enhances the efficiency of the industry due to its adaptive nature. On the other hand, sustainability refers to the fact that the technologies can be designed to be sustainable in nature and hence can be designed in a way such that these are both environment-friendly as well as eco-conscious in nature. This ensures that no harm is done to the environment as well as helps in reducing the overall energy as well as the metallic wastes produced by the industry.

7.1.2 Digital Twin

A DT is a virtual representation of a tangible system, activity, or component that steadily integrates real-time data from its real-world counterpart, ensuring dynamic bidirectional communication. This lets the digital version model, forecast, and enhance the performance of both itself and its physical version. The DT design enables the sharing of data in both directions between real and digital systems. This helps in smart monitoring, better analysis, and more informed decision-making.

The primary fact that differentiates a DT and a traditional simulation is the fact that traditional simulations are fixed and allow data to flow in only one direction. Unlike DTs, these older systems do not update or adapt based on real-time data once they are deployed. The concurrent data sharing between the real and digital systems enables predictive analysis, autonomous control, and system-level optimizations that traditional systems cannot achieve due to a lack of synchronization with the real-time systems. The detailed comparison between DTs and traditional simulators is illustrated in Table 7.2.

Table 7.2 Comparison of digital twins and traditional simulators.

Aspect	Digital twin	Simulators
Definition	A DT is a live, constantly updated digital version of a physical object or process, kept in sync with the original through continuous data streams and real-time updates	Static or semidynamic representation or simulation of a system that is based on historic data or predefined data without real-time synchronization
Real-time data integration	Synchronized in real time, hence is continuously updated in real time with the sensor data, IoT feeds, or telemetry or manual updates from the physical systems	No real-time data integration and synchronization, operates mostly on the predefined or manual data provided by the humans and requires periodic updates and redesigns
Level of synchronization	Maintains bidirectional synchronization, which means that any changes in the CPS reflect digitally in the DTs	One-way synchronization, which means the physical system impacts the model design but not vice versa
Purpose	Enables the twin to monitor, predict, and optimize, as well as provide necessary decisions across the lifecycle of the twin and the physical systems	Used for design and simulation at a specific interval of time or based on specific data
Lifecycle coverage	Extends across the life of the CPS from initialization phase till decommissioning	Often used only during design or testing phases and becomes obsolete postdeployment

(*Continued*)

Table 7.2 Comparison of digital twins and traditional simulators. (*Continued*)

Aspect	**Digital twin**	**Simulators**
Predictive capabilities	Uses machine learning and AI for real-time analytics to optimize performance, and enable prescriptive maintenance	Limited to basic simulations and what-if scenarios based on existing model data
Data handling and complexity	Heterogeneous data sources such as cloud, edge, and mist can be integrated into it, making design and data handling more complex	Uses and operates on a single dataset and data source and hence less complex to design and reduced complexity in handling the data because it is used for simulation purposes
Adaptability and learning	Uses feedback loops to constantly learn and adjust to situations and the behavior of the global model	Fully static and rule based in nature, hence requires manual tuning
Infrastructure requirement	Requires a complex and robust infrastructure such as cloud, edge, fog, real-time communication systems, etc., hence expensive in nature	Requires little infrastructure and is cheap because it can be simulated locally with minimal communication
Decision-making and control	Provides fully automated and semiautomated control with humans in loop, making it ideal for real-time systems	Humans make the decision based on the manual inputs and outputs
Scalability	Highly scalable, multiple twins can be combined to produce a digital thread	Limited scalability

(*Continued*)

Table 7.2 Comparison of digital twins and traditional simulators. (*Continued*)

Aspect	Digital twin	Simulators
Application areas	Widely used in smart manufacturing, healthcare, aerospace, energy systems, urban infrastructure, and federated IoT ecosystems	Mostly used in engineering design, testing environments, and academic modeling
Maintenance and cost efficiency	Reduces maintenance cost through predictive maintenance and operational optimization	Maintenance is reactive in nature, and keeping the model up-to-date with the physical systems is highly expensive and tedious in nature
Feedback mechanism	Deals with closed feedback loop	Deals with open feedback loops
Technological foundation	Built upon IoT, edge/cloud computing, data analytics, AI/machine learning (ML), principles of CPS and CPHS	Based on mathematical modeling, physics-based simulations, or static computation methods

7.1.3 Applications of DT

(i) Manufacturing

The DT can be used for enhancing the product standard and hence enhance the holistic production process and also predict equipment failures in the manufacturing industry. This helps in the enablement of the optimization of the product during design, predictive maintenance, and also improvement in the assembly or production pipeline. Applications also include real-time shop floor monitoring, simulated asset control, sustainability improvements, testing, and improved innovation, training, and space planning [6]. These features allow manufacturing firms to create a virtual copy and test and optimize the designs much before the actual physical production.

(ii) Construction

There are several challenges that are faced by the construction industry, and some of the key challenges include low productivity, a lack of research, and poor funding for innovation. The emergence and rapid advancement in DTs have aided in overcoming the aforementioned challenges, which are stated in [7]. In the construction industry, the DT enables predictive maintenance, which helps in the design and validation process of the physical systems.

(iii) Healthcare

Digital twins are used in healthcare as a replica or virtual double of the human body. In healthcare, the DT finds its use across various applications such as hospital management, product design for patients, system performance in the hospital, medical device design, innovation and research of new medicines, surgical planning, clinical trials, personalized medicine, and wellness, to list a few [8]. The DT in medical science provides personalized healthcare approaches by tailoring medication characteristics and treatment planning [9].

(iv) Smart Cities

The smart city is one of the applications in which the DT technology can be used very effectively. Smart maps and geospatial virtual infrastructures can be used to improve cities. Continuous monitoring of real-world aspects can be achieved through virtual models and machine learning (ML). The DT can be used as a testbed to perform experiments to validate multiple options. In urban planning, construction, transportation, and infrastructure development in cities [10], the DT is proved to be very efficient and economical. The key application subdomains and areas and the appropriate use cases of the DT with their key benefits are summarized in Table 7.3.

7.2 CPHS in the Present Era

CPHSs are modern systems with enhanced compute and communication powers to interact with humans by advanced means [11]. CPSs are considered to be a key part of automation in Industry 4.0 solutions. Integrating virtual models with their physical counterparts enables users to explore multiple "what-if" scenarios, conduct risk-free experimentation, and support predictive as well as proactive maintenance by forecasting future system states and identifying potential anomalies [12]. Many researchers

Table 7.3 Domain and application wise analysis of digital twins.

Domain	Subdomain/ area	Use case	Key benefit	References
Healthcare	Patient monitoring	Continuous vital-sign digital twin for early alerts	Early detection and personalization monitoring and treatment plans	[5, 8]
	Disease progression	Organ-level digital twin for disease simulation	Predictive diagnosis	[8]
	Personalized treatment/ drug testing	Virtual testing of therapies on patient digital twin	Safer, individualized therapy selection	[8]
	Medical equipment maintenance	Digital twin of imaging/ ventilators for faults	Reduced downtime	[5]
Manufacturing/ Industry 4.0	Production line monitoring	Machine-level digital twin for status and alarms	Reduced downtime	[23, 28]
	Process optimization	Workflow, energy, scheduling digital twin	Efficiency and energy savings	[28]
	Predictive maintenance	Asset digital twin predicting failures	Lower maintenance cost	[23, 28]
	Quality control	Digital twin–driven defect simulation	Better yield	[23]
Smart cities and infrastructure	Traffic and transport	City/road digital twin for congestion control	Reduced travel time	[7, 28]

(*Continued*)

Table 7.3 Domain and application wise analysis of digital twins. (*Continued*)

Domain	Subdomain/ area	Use case	Key benefit	References
	Building management (BMS)	Building digital twin for HVAC and energy	Energy optimization	[7]
	Urban planning	City-scale simulation digital twin	Policy testing and impact analysis	[7]
Aerospace	Aircraft health monitoring	Engine and component digital twin	Improved safety	[23, 24]
	Mission/flight simulation	Flight digital twin for mission planning	Risk reduction by optimizing the flight mission *via* simulation	[24]
Automotive	Vehicle dynamics and ADAS	Vehicle/ environment digital twin for testing	Safer control and validation	[23, 28]
	Battery management	Battery digital twin for SOH/SOE prediction	Extended battery life; better scheduling	[28]
Energy and power systems	Grid monitoring	Grid node digital twin for fault detection	Faster fault isolation; stability	[28]
	Renewables (turbines, PV)	Turbine/plant digital twin for performance	Increased energy yield	[5, 28]
Construction and civil	Structural health	Bridge/building digital twin for monitoring	Early damage detection	[7]

(*Continued*)

Table 7.3 Domain and application wise analysis of digital twins. (*Continued*)

Domain	Subdomain/ area	Use case	Key benefit	References
	Construction logistics	Digital twin for resource and schedule tracking	Reduced delays	[7]
Agriculture	Precision farming	Crop/soil digital twin for irrigation and input	Yield increase	[28]
	Machinery monitoring	Tractor/drone digital twin for maintenance	Lower downtime	[28]
Maritime/ offshore	Vessel condition monitoring	Ship digital twin for propulsion and hull	Preventive Maintenance and fuel savings	[23, 24]
Logistics and supply chain	Warehouse/ route digital twin	Warehouse twin and route simulation	Faster throughput	[23, 28]
Education and research	Virtual labs and training	digital twin simulators for hands-on practice	Safe, repeatable training	[5, 24]
Cybersecurity/ CPS	Threat simulation and resilience	Digital twin of industrial control systems for attacks	Safe security testing	[24, 28]

throughout the world have successfully implemented DTs in diverse fields to have risk-free testing activities.

DT technology can be used in smart manufacturing for threat detection. The DT system simulates the threats and detects them using deep learning–based time-series classification models. The DT testbed was built using Factory I/O simulation software and the OpenPLC platform [13]. This CPS setup processes in smart manufacturing, representing sensors, actuators, and control logic. The final testbed generated is cost-effective, easy to replicate, and easy to modify. Various threat scenarios were simulated, mostly focusing on the speed and direction of conveyor belts in an automated quality checking process. A variety of deep learning models for time-series classification were trained and evaluated. The DT technology has provided the solution to the data-scarcity issues.

The DT framework helps in the designing, implementing, and functioning of cyberphysical infrastructure. The framework maps the fundamental elements of DTs, such as data, process models, and real-time synchronization, into collaborative environments to interact dynamically. The architecture provides bidirectional data flows between the CPSs and the DTs, and collaboration protocols for interactions and decision-making. This framework analyzes the use cases to highlight the collaborative and adaptive functionalities provided by the DTs. Digital twins and CPS systems are expected to drive automation and enable real-time monitoring [14]. The CPSs connect with their digital copies using sensors and other Internet of Things (IoT) devices. This framework elaborates the use cases of DT technology for engineering and infrastructure development. CPS-based IoT architecture has the potential gains in workplace productivity, health and safety, building management, and workforce skills enabled by these systems. The grounded theory approach analyzes complex, interconnected systems and highlights mechanisms driving artificial intelligence (AI) development in IoT networks through the study of human–computer interactions in information management systems [15].

Air-gapped systems, a lack of standardization, an increased surface attack area, insider threats, and supply chain vulnerabilities are some of the issues faced by modern nuclear reactors while communicating with modern technologies for cybersecurity needs. They are connected with the smart grid, monitored remotely, and run efficiently. This requires secure two-way communication [16]. This communication exposes the reactors to cyber threats, so very strong cybersecurity measures are required. The study proposes a DT-based testbed at Purdue University's PUR-1 reactor. The reactor uses sensors, actuators, and an auxiliary moderator displacement rod to mimic the reactor operations from remote locations. Simulated cyberattacks were run and compared testbed data with real PUR-1 data, using physics-based models to analyze the behavior beyond direct measurements.

There are several approaches for testing CPS systems, including model-driven, heuristic-driven, data-driven, monitoring-driven, fault-simulated, and cloud-driven methods. The model-driven approach is used for software testing. Search-based testing uses metaheuristic search methods to generate the test data. The monitoring method is the runtime verification of observing the real-time behavior or the simulation. Fault injection is a technique where failures are deliberately introduced to accelerate the detection or occurrence of errors in a system [17]. The big data–driven testing method provides a mechanism to couple big data and CPSs. Most applications utilize DTs as testbeds, where this hybrid approach allows the DT to adapt, analyze, and mirror the CPS's behavior under various conditions, enabling risk-free decision-making, predictive maintenance, and effective planning. The next section deals with a more detailed analysis of DT testbeds and their architecture and applications across various domains.

7.3 Digital Twin Test Beds

A DT testbed uses a hybrid modeling and validation environment that integrates the physical, virtual, and data-driven components in the DT to replicate the scenarios and system behaviors of the physical components, under a set of controlled and measurable conditions [18]. The DT testbed combines multiple modeling techniques to generate an accurate system representation. It uses physics-based models to simulate the fundamental system behavior and ensures that the physical behavior is reproduced accurately. The hard-to-capture behaviors, such as unmodeled and nonlinear behaviors, are modeled using data-driven models. Real-time and historical datasets are used to improve the predictive capability. A hybrid fusion framework combines the two paradigms. ML algorithms augment the physical models, improving robustness, adaptability, and accuracy under varying conditions. Extended Kalman filters (EKFs) perform real-time state estimation. They continuously correct model predictions with sensor measurements, maintaining synchronization between the simulated and actual system states. The key comparison between a DT, a DT testbed, and a normal testbed is illustrated in a detailed manner in Table 7.4.

The above table showcases the key differences between DT testbeds and traditional testbeds in a detailed manner. The architecture of the DT testbed is discussed in a more detailed manner in the section below.

7.3.1 Architecture of Digital Twin Testbed

The DT testbed architecture has five functional layers, each designed to distribute responsibilities and simplify operations (Figure 7.1).

(i) Physical layer: It comprises all the real-life settings and devices, such as industrial equipment, automotive, and healthcare equipment. These are

Table 7.4 Domain and application wise analysis of digital twins.

Aspect	Digital twin	Digital twin testbed	Normal testbed
Purpose	Represents and monitors a physical system in real time	Validate, test, and refine the digital twin models under realistic hybrid scenarios	Test the performance of a component or system under a simulated and controlled laboratory environment
Nature	Virtual and data-driven	Hybrid	It may exist as a physical or simulated model; however, it is not connected or refreshed in real time
Connection to real system	Always connected to a live physical system	Connected to both physical and virtual systems so as to enable hybridity	Isolated from the real-time systems and usually not up-to-date with the existing physical systems
Data flow	Continuous bidirectional flow due to bidirectional synchronization between the CPS system and the DT	Controlled unidirectional and bidirectional communication during simulation and updation, respectively	Single time data capture to simulate the outcome based on the specified conditionals and control variables
Focus area	Real-time operation, prediction, and optimization	Experimentation, model validation, and synchronization testing	Basic performance testing or hardware verification

(*Continued*)

Table 7.4 Domain and application wise analysis of digital twins. (*Continued*)

Aspect	**Digital twin**	**Digital twin testbed**	**Normal testbed**
Level of integration	Deep integration with IoT, AI, and cloud systems	Partial integration due to its hybrid setup for experimentation	Limited or no integration with smart or connected systems
Real-time capability	Operates either entirely in real time or with minimal delay	Semi–real-time communication based on the test simulations done as well as the testbed setup	Often offline or postprocessing based
Use of simulation	Simulation is embedded as part of continuous digital modeling	Simulation used to validate or tune the digital twin	Simulation used independently for isolated experiments
Feedback control	Provides live feedback and control to the physical asset	Semi feedback control can offer both semiopen and semiclosed loops	No feedback control, manual tuning is needed
Scalability	Scalable across large systems or networks	Limited to specific environments and use case	Less scalable in nature
Flexibility	Highly adaptive to system changes in real time	Adaptive but may require reconfiguration in some scenarios	Fixed setup for predefined tests
Maintenance requirement	Requires continuous data updates and recalibration	Requires regular validation experiments	Minimal, requires very little maintenance

(*Continued*)

Table 7.4 Domain and application wise analysis of digital twins. (*Continued*)

Aspect	Digital twin	Digital twin testbed	Normal testbed
Typical domains	Smart manufacturing, smart cities, healthcare, aerospace	Research labs, pilot factories, system design validation setups	Hardware testing, embedded systems, sensor validation
Cost factor	High due to data infrastructure and computation	Moderate to high—depends on hybrid setup	Lower—standard laboratory equipment-based

integrated with sensors and actuators so as to measure key changes and operational parameters and hence induce the required change in the twin or the environment based on the feedback command to be executed. All the data are sent from this layer to the DT, and hence, this layer can be considered to be the data source for all the remaining layers.

(ii) Cyber layer: The cyber layer includes the required embedded software and the control logic for process regulation and synchronization. The cyber layer provides an interface for the DT and the CPS to enable the sharing, processing, and optimization of the information shared between them. The cyber layer executes the required simulation and analytics and performs the process control at the virtual and the actual systems.

(iii) DT layer: This layer, comprising the DT models, is composed of high-precision digital models of real-life settings and systems that are synchronized with the physical system. The DT layer is always in a bidirectional synchronization with the tangible layer with the help of the synchronization layer.

(iv) Synchronization layer: This facilitates maintaining the real-time or near-real-time consistency among the CPS and the DT layer. The data flow, update frequency, state consistency, and bidirectional communication are handled in this layer.

(v) Experimental layer: The experimental layer allows for testing various configurations on the DT *via* simulation and validation before these are applied to the real-world physical systems. This layer can be considered a sandboxed or isolated environment where the testing and validation of various hypotheses can be conducted without affecting the real-world systems.

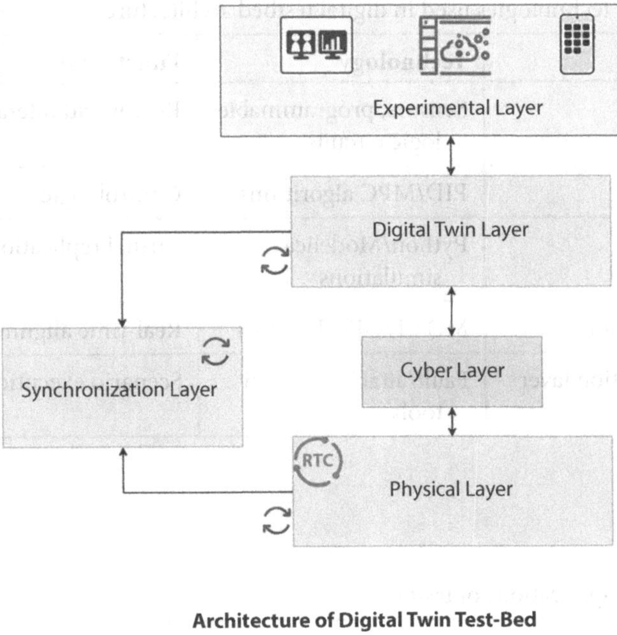

Figure 7.1 Architecture of digital twin testbed.

Figure 7.1 illustrates the detailed architecture of the DT testbed. The set of communication protocols used in DT testbeds is discussed in a more detailed manner in the next section of this article.

7.3.2 Communication Protocols in Testbeds

Within the testbed environment, an effective, secure, and compatible data-sharing environment is provided by the communication framework; it makes use of message queuing telemetry transport (MQTT). This protocol is explicitly structured to ensure seamless communication in networks characterized by limited bandwidth and significant latency. MQTT enables smooth instant data sharing between distributed components and the central server, while keeping communication overhead low [19]. To enhance industrial compatibility, the framework utilizes open platform communications unified architecture (OPC UA). This provides a standardized, platform-agnostic interface that supports structured data modeling, secure information exchange, and connectivity among various industrial systems [20].

Table 7.5 Key technologies used in digital testbed architecture.

Layer	Technology	Functionality
Physical	Sensors, programmable logic circuits	Real-world interaction
Cyber	PID/MPC algorithms	Control logic
Digital twin	Python/Modelica simulations	Virtual replication
Synchronization	MQTT/OPC UA, EKF	Real-time alignment
Experimentation layer	Fault/attack injection tools	Scenario execution

Table 7.6 Key applications of testbeds.

Domain	Digital twin testbed use case	Purpose	References
Smart manufacturing	Robotic assembly line testbed	Predictive maintenance and process optimization	[23, 25]
Healthcare	Patient-monitoring IoT testbed	Real-time health analytics and anomaly detection	[24, 25]
Smart grid	Distributed energy network testbed	Load balancing and fault prediction	[25, 27]
Transportation	Autonomous vehicle control testbed	Safety and reliability testing	[26, 27]
Aerospace	Flight simulation testbed	Component fatigue and predictive fault analysis	[26, 27]

All time-series data produced by sensors, controllers, and simulations are stored in InfluxDB, a high-performance time-series database optimized for quick data ingestion and retrieval. Grafana is used for data visualization and monitoring [21, 22]. It offers dynamic dashboards and analytical tools that help users track system performance, identify anomalies, and visualize trends in real time. Communication security and data integrity are maintained through transport layer security encryption across all communication channels. Role-based access control mechanisms limit unauthorized access and protect the confidentiality of exchanged information. Table 7.5 illustrates the tech stack for communication in a testbed setting in a detailed manner. Similarly, the set of applications of the testbed can be seen in Table 7.6.

7.4 Experimental Scenarios

The section below introduces various operating scenarios of DT and CPS testbeds.

7.4.1 Controller Tuning

Two control strategies were implemented and evaluated: the model predictive control (MPC) and proportional integral derivative (PID) controllers. Using DTs, both controllers were assessed for the corresponding physical system. The main goal was to identify which controller maintains system stability and optimizes the dynamic performance. The key parameters used are settling time, overshoot, and energy consumption.

Due to its simplicity, the PID controller is most widely used in industrial automation. It has shown greater overshoot and slower settling times under dynamic conditions. On the other hand, the MPC controller used model-based predictive optimization. This improves the system variations and optimization of control inputs. This approach resulted in improved settling time, reduced overshoot, and increased energy efficiency, consistent with previous findings.

7.4.2 Fault Injection

Fault injection experiments can be carried out on the DT so as to assess the system's robustness and its fault tolerance. The DT, along with its testbed, enables detailed simulations of sensor malfunctions, delays in actuators, and loss of communication without altering or compromising the physical real-time system. This helps to evaluate and understand how the system responds under potential conditions that may lead to its failure.

Some of the most commonly used fault scenarios used in DT testbeds are signal drift, packet loss, and timing jitters in the networks, which can be used to analyze the impact of their effect on the stability of the system. The testbed's ability to replicate various realistic operational conditions ensures repeatability and allows for safety during experimentation of fault injection in the system.

7.4.3 Cybersecurity Testing

Cybersecurity validation in DT testbeds focuses on the detection and mitigation of various cybersecurity threats across the physical and DT layers in real time. The DT enables a secure and regulated setting that can be used to simulate and emulate various cybersecurity attacks such as denial-of-service, data spoofing, and botnet-based command injection on the CPS network. The testbed allows cybersecurity teams to reproduce a set of cyber events and scenarios virtually. This enables the cybersecurity teams to build robust security measures and simulate the set of attacks in real time, without impacting its performance and operation. The DTs analyze the logs and run several integrity checks on the existing system and hence can help in detection and remediation when under cyberattack.

7.4.4 Model Calibration

Model calibration ensures that the virtual copy behaves exactly as the modeled physical system. The calibration can be performed on the DT by adjusting the sensor data and other system outputs with the efficient values predicted by the DT. Several key parameters such as transfer functions, gain constants, and dynamic coefficients can be changed iteratively until convergence or the desired result is achieved. The synchronization layer of the testbed enables several real-time parameter updates and hence increases the accuracy of the DT.

7.4.5 Resilience Assessment

Resilience refers to the system's capability to resist any abnormal conditions that may affect its normal working. Resilience assessment is the capability of the physical system to adapt and overcome any unexpected disturbances while ensuring that its operation and its performance are not impacted. Using DTs, several disturbances—such as component failures, sensor failures, load variations for scalability tests, and cyberphysical attacks—can be introduced to the virtual twin on the testbed and its recovery process and resiliency can be monitored.

7.5 Conclusion

The architecture for the DT-based testbed demonstrates the real-time integration among the CPS and its virtual counterpart to provide an effective testing environment. The layered architecture distributes the responsibility across the multiple layers and assigns distinct roles to each layer. Conceptually, this architecture is expected to provide bidirectional communication, predictive analysis, and safe maintenance without interrupting the regular operations of the CPS. The five-layered framework of the DT testbed promises to provide scalability, interoperability, and flexibility, which makes it suitable for applications such as manufacturing, construction, healthcare, etc. The DT testbeds are said to be advantageous when used for CPSs. The five-layered DT architecture provides advantages in areas such as monitoring and control, predictive maintenance, data-driven optimization, scalability, lifecycle management, enhanced safety and risk mitigation, a controlled experimentation environment, hybrid validation, accelerated innovation, risk-free testing, and personalization. Similarly, some major disadvantages include high cost, complex design and modeling, synchronization issues, security challenges, high costs of implementation, limited scalability for complex models, maintenance overhead, and data-integration complexity. This DT testbed architecture offers a systematic and modular framework that integrates both physical and virtual components of a CPS. The synchronization layer ensures data consistency, whereas the experimental layer facilitates analysis, algorithm validation, and virtual testing across a range of scenarios.

7.6 Future Scope

The DT application is not restricted to the aforementioned; it can also be explored for the following areas in future research works. Table 7.7 showcases the set of future work and its impact on the existing DT technologies.

Table 7.7 Research areas and future works in digital twins.

Research area	Future work area	Impact
Interoperability and standards	Develop standard and unified data exchange protocols	Enables seamless integration across various heterogeneous systems
AI-driven twins	Integration of various adaptive AI/ML models for autonomous decision-making in a digital twin environment	Improvement in real-time accuracy and adaptability
Cybersecurity and privacy	Secure and reliable synchronization techniques, encryption pipelines for the whole process, and enablement of anomaly detection in digital twin networks	Increases the overall trust and usage of digital twins across industries
Scalability and edge deployment	Enable light weight digital twins by using edge devices and enable federated machine learning if needed	Reduces latency in the network and enables decentralized machine learning
Hybrid modeling approaches	Combine physics-based and data-driven models in a more efficient way	Increases model fidelity under dynamic conditions
Human–twin interaction	Develop a robust and user friendly interface to interact and visualize the digital twin	Enables interoperability and decision support in the digital twin environment

(*Continued*)

Table 7.7 Research areas and future works in digital twins. (*Continued*)

Research area	Future work area	Impact
Cross-domain integration	Multidomain digital twins combining the physical systems across various domains	Enables city-wise integration and can be used in smart city applications
Validation and benchmarking	Establish common testbeds for digital twins to compare standard performance and hence increase the quality of service of the digital twin system	Enables consistency and enhances its application across research applications

AI disclosure statement: The author confirms that no artificial intelligence (AI)-generated content or AI-assisted writing tools were used in the preparation of this chapter (including summaries and the abstract). All text, analysis, and conclusions are the result of the author's original work.

References

1. Pasupuleti, M.K., Revolutionizing Industries with Digital Twin Technology. *Digi. Twin Tec.*, 12, 61, 2024.
2. Kanaga Priya, P. and Reethika, A., A Review of Digital Twin Applications in Various Sectors. *Trans. Ind. Using Digi. Twin. Tech.*, 5, 219, 2024.
3. Latorre, L., Rego, E., De Leo, L., Gutierrez, M., Digital Twins. *Inter-American Dev. Bank. Tech. Report*, 3, 136, 2024.
4. Mythily, M., David, B., Vijay, J.A., Digital Twin Application in Various Sectors. 2024. in: *Proc. of Transforming Indu. Using Digi. Twin. Tech.*, 3, 219, 2024.
5. Sharma, A., Kosasih, E., Zhang, J., Brintrup, A., Calinescu, A., Digital Twins: State of the art theory and practice, challenges, and open research questions. *J. Ind. Inf. Integr.*, 30, 100383, 2022.
6. Jyothi, A.P., Shankar, A., Narayan, A.J.R., Koliparthy, K., Badiger, V.N., Shivakumar, V., Exploring the Applications and Challenges of Digital Twins in Various Industries. in *2024. Int. Conf. Intell. Comput. Technol.*, 4, 8, 2024.

7. Opoku, D.G.J., Perera, S., Osei-Kyei, R., Rashidi, M., Digital twin application in the construction industry: A literature review. *J. Build. Eng.*, 40, 102726, 2021.
8. Katsoulakis, E., Wang, Q., Wu, H., Shahriyari, L., Fletcher, R., Liu, J., Deng, J., Digital twins for health: a scoping review. *npj Digit. Med.*, 7, 77, 2024.
9. Jyothi, A.P., Shankar, A., Narayan, A.J.R., Koliparthy, K., Badiger, V.N., Shivakumar, V., Exploring the Applications and Challenges of Digital Twins in Various Industries. in: *2024. Int. Conf. Intell. Comput. Technol.*, 1, 8, 2024.
10. Halúsková, B., Digital twin in smart city. *Transp. Res. Procedia*, 74, 1471, 2023.
11. Baheti, R. and Gill, H., Cyber-physical systems. *Impact Control Technol.*, 12, 161, 2011.
12. Lo, C., Win, T.Y., Rezaeifar, Z., Khan, Z., Legg, P., Digital Twins of Cyber Physical Systems in Smart Manufacturing for Threat Simulation and Detection with Deep Learning for Time Series Classification. in: *Proc. 29th Int. Conf. Autom. Comput. (ICAC)*, 1, 6, 2024.
13. Nazarenko, A.A. and Camarinha-Matos, L.M., The role of digital twins in collaborative cyber-physical systems. in: *Doctoral. Conf. Comput. Electr. Ind. Syst.*, 4, 191, 2020.
14. Akanmu, A.A., Anumba, C.J., Ogunseiju, O.O., Towards next generation cyber-physical systems and digital twins for construction. *J. Inf. Technol. Constr.*, 26, 2021.
15. Radanliev, P., De Roure, D., Nicolescu, R., Huth, M., Santos, O., Digital twins: artificial intelligence and the IoT cyber-physical systems in Industry 4.0. *Int. J. Intell. Robot. Appl.*, 6, 171, 2022.
16. Theos, V., Design and development of a real-time cyber-physical testbed for cybersecurity research, Doctoral dissertation, Purdue University Graduate School, 2023.
17. Zhou, X., Gou, X., Huang, T., Yang, S., Review on testing of cyber physical systems: Methods and testbeds. *IEEE Access*, 6, 52179, 2018.
18. Harper, K.E., Malakuti, S., Ganz, C., Digital twin architecture and standards, 2019.
19. Wang, Z., Xiao, H., Guan, C., Zhou, L., Fu, D., Research on the Development of a Building Model Management System Integrating MQTT Sensing. *Sensors*, 25, 6069, 2025.
20. Leitner, S.H. and Mahnke, W., OPC UA–service-oriented architecture for industrial applications. *Softw. Trends*, 26, 2006.
21. Kagermann, H. and Wahlster, W., Ten years of Industrie 4.0. *Sci.*, 4, 26, 2022.
22. Ivanov, D., The Industry 5.0 framework: viability-based integration of the resilience, sustainability, and human-centricity perspectives. *Int. J. Prod. Res.*, 61, 1683, 2023.
23. Kritzinger, W., Karner, M., Traar, G., Henjes, J., Sihn, W., Digital Twin in manufacturing: A categorical literature review and classification. *IFAC-Pap. Online*, 51, 1016, 2018.

24. Fuller, A., Fan, Z., Day, C., Barlow, C., Digital Twin: Enabling technologies, challenges and open research. *IEEE Access*, 8, 108952, 2020.
25. Grieves, M. and Vickers, J., Digital twin: Mitigating unpredictable, undesirable emergent behavior in complex systems. in: *Transdisciplinary perspectives on complex systems: New findings and approaches*, 12, 85–113, Springer International Publishing, Cham, 2016.
26. Madni, A.M., Madni, C.C., Lucero, S.D., Leveraging digital twin technology in model-based systems engineering. *Syst.*, 7, 7, 2019.
27. Tao, F. and Zhang, M., Digital Twin Shop-Floor: A New Shop-Floor Paradigm Towards Smart Manufacturing. *IEEE Access*, 5, 20418, 2017.
28. Qi, Q. and Tao, F., Digital twin and big data towards smart manufacturing and industry 4.0: 360 degree comparison. *IEEE Access*, 6, 3585, 2018.

8

A Holistic Review of Ethical, Legal, and Social Implications of Emerging Cyberphysical Human Systems

Bhuvana B. P.[1], Bharathi Raju[1*], V. Yokesh[1] and Ahmed A. Elngar[2]

[1]Department of Electronics and Communication Engineering, School of Electrical and Electronics, Sathyabama Institute of Science and Technology, Chennai, Tamil Nadu, India

[2]Computer Science Department, Faculty of Computers and Artificial Intelligence, Beni-Suef University, Beni Suef, Beni Suef Governorate, Egypt

Abstract

The future systems, which integrate computers, real-world processes, and people, are cyberphysical human systems (CPHSs), which allow smart and adaptable interactions between the systems and people. CPHSs, despite their revolutionary potential in the fields of healthcare, autonomous transport, intelligent environments, and manufacturing, have difficult ethical, legal, and social implications. In this chapter, these implications are explored in more detail, and it is clear that privacy, security, transparency, and responsibility are under urgent consideration. CPHSs have ethical concerns regarding the issue of informed consent, the ownership of data, and predictive biases that can result in discrimination or exclusion. As a matter of law, the current rules and regulations cannot keep up with the fast-changing technology transitions, leaving loopholes in the areas of liability, intellectual property rights, and cross-border data governance. Whenever CPHSs are in use, they affect human autonomy, trust, and societal norms, and may contribute to the expansion of digital divides and remodeling labor markets. These challenges are critically examined in this paper, and the emphasis on the need for the technical design to be balanced with human-centered values is illuminated. It also demonstrates the need to have an interdisciplinary cooperation to come up with strong rules and procedures that make sure that CPHSs are applied fairly and

**Corresponding author*: bharathiraju90@gmail.com

Anitha Velu, Prasanth Aruchamy, Raghu Ramamoorthy, Rajesh Kumar Dhanaraj and Manjula Arunraj (eds.) Cyber Physical Human Systems: Emerging Trends and Applications, (179–208) © 2026 Scrivener Publishing LLC

in a responsible manner. The primary purpose of the chapter is to contribute to the current discussion on enhancing CPHSs through the reconciliation of innovation and societal requirements.

Keywords: CPHS, security, ethical design, legal consideration, social implications

8.1 Introduction

Cyberphysical human systems (CPHSs) are used today to bring together people, digital technology, and physical plants with the aim of achieving a goal. Cyberphysical systems (CPSs) need people to connect networking, computing, governance, and connectivity into physical systems. People are considered an important part of these systems. As problems in society grow, people work with CPSs to solve the many and complicated problems they face. As CPSs become more common, they change how people interact with smart computer systems. CPSs make automation better by using artificial intelligence in open and changing environments [1]. For instance, in the healthcare field, doctors give patients medicine based on information gathered by portable medical tracking devices. Similarly, those who own smart homes can control their smart devices using voice commands or mobile apps [2].

Due to the crucial role of CPSs in numerous essential infrastructures, such as transportation, energy, healthcare, and smart cities, their influence transcends mere technological functionalities. Although they provide undeniable benefits for efficiency, responsiveness, and data-informed decision-making, they also present significant issues that extend beyond engineering, such as ethical quandaries, legal uncertainties, and societal disturbances. These apprehensions arise from the close interconnection of data acquisition, automated decision-making, and tangible acts.

8.2 Ethical Consideration for CPS

Emerging technologies have evolved to a greater extent in that they rely upon software platforms. However, if a platform is exploited, then the information that has been retrieved can be manipulated or stolen, leading to privacy breaches, and causing physical, mental, and financial harm to the users. Hence, the ethical consequence of the CPS is significant as the systems directly affect human safety, autonomy, and privacy, as well as equity. Ethical considerations in CPSs show a clearly organized picture of

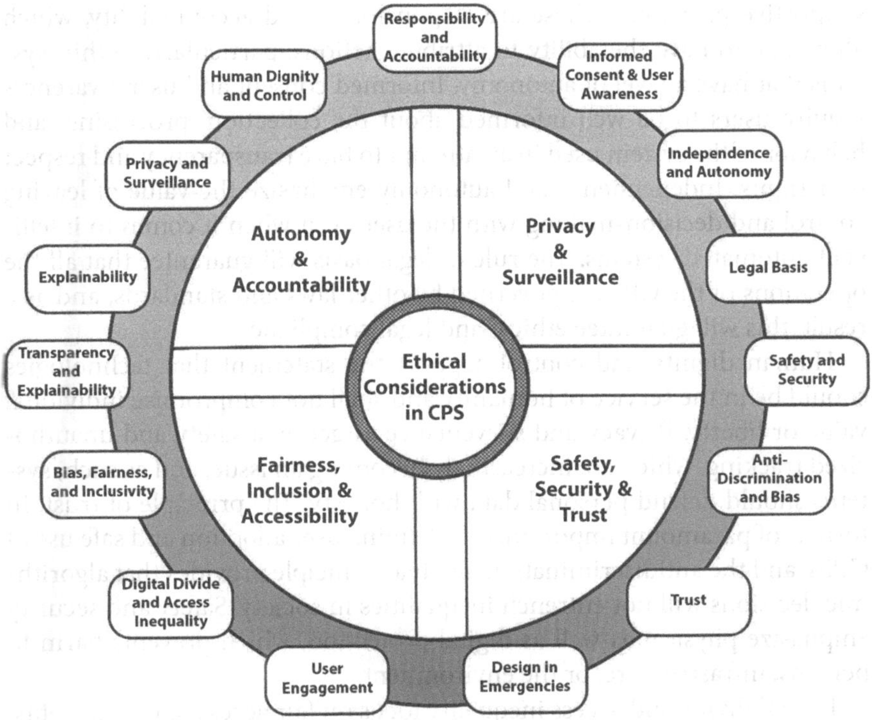

Figure 8.1 Ethical considerations in CPS.

the vital sets of ethical areas and concepts needed to design and oversee CPSs, as depicted in Figure 8.1.

A few theme areas characterize the CPS, namely autonomy and accountability; privacy and surveillance; fairness, inclusion, and accessibility; and safety, security, and trust. Autonomy and accountability dictate the importance of appreciating the element of user independence, but there should be established mechanisms for tracing decisions made within the system and taking ownership of the decisions. Privacy and surveillance touch upon the issues of data gathering, observation, and possible manipulation, supporting reasonable data management. Fairness, inclusion, and accessibility mean that CPS technologies are not made with a specific user group in mind, are designed to be inclusive of a wide variety of other users, and reduce digital divides without increasing them. The principles of safety, security, and trust emphasize the significance of creating physically and digitally secure systems that inspire confidence in the user.

The presented diagram offers a detailed understanding of the ethical consequences of the CPS by pointing out a variety of fundamental and supportive principles. These are responsibility and accountability, which allow one to have the ability to attribute actions, particularly within systems that have a level of autonomy. Informed consent and user awareness require users to be well informed about the collection, processing, and behavior of the system used in an attempt to have transparency and respect user rights. Independence and autonomy emphasize the value of leaving control and decision-making with the user even when it comes to intelligent automated systems. The rule of legal basis will guarantee that all the operations of the CPS are governed by other laws and standards, and as a result, this will guarantee ethical and legal compliance.

Human dignity and control support the statement that technologies should be in the service of humanity and shall not compromise individual value or liberty. Privacy and surveillance target data safety and unauthorized tracking, which are increasingly becoming an issue, and as such, systems should defend personal data with honesty. The principle of trust, in turn, is of paramount importance in defining user adoption and safe use of CPSs, and the antidiscrimination and bias principle provides that algorithmic decisions will not entrench inequalities in society. Safety and security emphasize physical as well as digital protection, which prevents harm to persons, infrastructure, or the environment.

Digital divide and access inequality focus on fair access to CPS benefits, particularly to marginalized populations, and user engagement focuses on inclusive design by including stakeholders in all phases of the system life cycle. Design in emergencies requires ethical responsive behavior under situations of crisis where a system may be able to change its structure and operate robustly with a sufficient degree of certainty. Finally, transparency and explainability mean that the choices made by the CPS should be clear and easy to understand. The fact that users can hold the CPS responsible and have confidence in it is crucial. These interrelated aspects constitute a holistic approach to an ethically informed approach to planning, implementing, and managing CPSs that leads to transparent, equitable, secure, and ethical systems.

8.2.1 Autonomy and Accountability

There are huge moral concerns surrounding CPSs, such as accountability and autonomy. Autonomy implies that a system can act and make its choices without real human control, as autonomous cars demonstrate or robotic healthcare systems. This adds efficiency and responsiveness as well as posing severe ethical concerns if a system makes malicious or malfunctioning decisions. Next arises the question of accountability: Is it the developer, the manufacturer, the user, or the AI itself? Such confusion compromises older conceptions of responsibility and obligation.

The above challenges fall within a five-level CPS (5C) model that subdivides CPS capabilities into five horizontal levels: smart connection, data-to-information conversion, cyber, cognition, and configuration. Actuators and sensors provide direct contact with the actual world to gather raw data on the smart connection level. This layer is almost independent, increasing responsiveness and decentralizing the sense of control, making it more difficult to find issues and place responsibility.

The data-to-information conversion layer converts raw data into an understandable format; this is through filtering, aggregation, and signal interpretation. This is necessary for cognition at the higher level, yet should it be done without proper control, it may distort data, or context can be lost. The cyber layer provides secure communication, exchange of data, and interoperability between distributed CPS components. It gives informational exchange a basis and is emerging as one of the main areas of concern regarding cybersecurity, such as data privacy, reliability, and accessibility threats.

The cognition layer is an assessment of subjective data performed to enhance decision-making by rule-based algorithms, pattern recognition, or artificial intelligence. This layer demands clear and transparent decision-making frameworks to enable autonomous actions to be tracked and be answerable. The configuration layer depicts the human–machine interface, which introduces human supervision into the CPS decision-making process. Human-in-the-loop control is required in critical applications, such as autonomous driving or robotic-assisted surgery, where an ethical challenge is imposed on the automated decision, and where the decision may need to be overridden.

The configuration layer acts as a very important checkpoint where the judgment of a human over machine autonomy may occur where necessary. The 5C architecture enhances the efficiency of CPS activity and provides a systematic mechanism for placing ethical safeguards. A balance between physical data collection and decision-making in the human will promote public trust and adherence to ethical standards. A view of autonomy and tasks in CPSs is given in Table 8.1.

8.2.2 Privacy and Surveillance

The two most noticeable ethical issues of CPSs are privacy and monitoring. These systems gather much personal and environmental information, such as locations, health measurements, behavioral history, and private

Table 8.1 5C framework of CPS autonomy and accountability.

Sl. no.	5C layer	Autonomy role	Accountability mechanism	Ethical implication
1.	Configure	Human in the loop decision authority	Supervisory control, override functions, and ethical configuration logs	Human oversight for critical control
2.	Cognition	AI-driven decision-making and optimization	Explainable AI, decision traceability, model validation	Transparent and auditable decision paths
3.	Cyber	Distributed system coordination, real time communication	Secure logging, peer auditing, identity verification	Responsibility in data flow and system behavior
4.	Conversion	Semiautonomous data processing and diagnostics	Anomaly reporting, performance logging, fault localization	System self-awareness with logged activities
5.	Connection	Autonomous data collection by edge devices/ sensors	Sensor certification, integrity checks, tamper detection	Ensures data trustworthiness at the edge

discussions, and apply different sensors, cameras, and activity devices. Such information simplifies and speeds up systems, but it also poses a risk to the privacy of the user. Crimes such as unauthorized access to data, hacking, or spying without express permission go against basic rights and liberty.

Permanent observation of public and private spaces produces a feeling of being surveilled. This may cause behavioral alteration, mental torture, and deprive people of their freedom. To take an example, smart-city monitoring systems can be run without the knowledge or permission of people, casting doubts on the notion of informed consent and democratic control [3]. Such systems are also problematic around the matter of ownership of data and future applications. A variety of users have no idea who accesses the data collected by the CPS, how they can be used to target advertising or to evaluate insurance risk, or even their use by law enforcement without user approval.

These issues lead to "function creep," whereby information obtained to use for one purpose is applied to another. To solve these issues, the CPS has to be designed and operated by clearly defined ethical principles [4]. There are central practices, such as informed consent, collecting only the necessary data, and anonymizing them so that they cannot be identified. The most important thing is algorithmic fairness, particularly in technology capable of determining decisions about people, e.g., predictive policing, healthcare diagnostics, or resource distribution. Discrimination and marginalization of underprivileged communities can be caused by bias in data or algorithms, entangling ethical questions. The conclusion is that there has to be a middle ground between technological advancement and privacy rights. Individuals also rely on the CPS due to the belief that it not only is technically reliable but is considered to carry moral and social norms. With the help of their ethical design, governmental supervision, and engagement of the public, the CPS should be socially acceptable and responsible.

8.2.3 Safety and Security

The CPS is a critical ethical issue when it comes to designing and implementing safety and security. CPSs have the potential to cause physical harm, including injury, financial loss, or even loss of life because, by touching the physical world, any failure in their functioning or protection can cause physical harm [5]. An illustration is given, i.e., a hacked autonomous car or tampered medical equipment may lead to disastrous consequences. Keeping the system safe would imply that the system will be capable of working properly even when subjected to all anticipated faults, whereas security would entail defending it against cyberattacks and illegal access and use. The CPS should be well-architected in such a way that it has good encryption, access controls, and is monitored in real time. Both aspects are instrumental in building the trust of the people, and any lapse in them can lead to wide disruption [6]. It is an ethical responsibility of developers to foresee what might go wrong and develop a fail-safe, robust system. Routine testing and updating have to be done to prevent threats. Finally, safety and security come first, as no one can choose them as a secondary parameter of CPS devotion.

8.2.4 Bias, Fairness, and Inclusivity

The development of the CPS is sensitive to several ethical aspects that include bias, fairness, and inclusivity. Such systems are frequently backed by artificial intelligence and machine learning that unintentionally reproduce the biases that are in place in their training data. Such biases may result in unfair or discriminatory consequences unless they are mitigated, including, but not limited to, supporting the interests of some groups over others in medical care, recruitment, or monitoring initiatives. An example is that a facial recognition algorithm can be biased against those with darker skin because of the bias of the datasets against them [7].

Also, fairness implies that the CPS has to treat all users equally and has to give them equal opportunities, irrespective of gender, race, age, and background. The inclusivity aspect should involve a variety of ideas in the designing and testing processes in order not to leave out minority needs. Fairness audits and transparency of decision-making should be part of the ethical development of the CPS. Methods of reducing biases must be used in the life cycle of the system as a whole [8]. An inclusive and equitable CPS attracts social justice and a sense of trust in technology.

8.2.5 Human Dignity and Control

The major ethical considerations when developing CPSs are respect toward individuals and control. Human intervention on important issues such as healthcare, law enforcement, or caregiving may be eliminated (or overridden) by these independent systems becoming more automated, thereby replacing or entirely removing human intervention. This may undermine individual agency and depersonalize the individual, making him or her a data point. Human dignity plays a significant role in treating people with respect to their rights, values, and emotional needs in dealing with the CPS [9]. People should always be in control, particularly in decisions that affect their lives. The CPS is made to support and enhance human capabilities but not to dominate and replace human beings. Oversight and accountability are maintained with such features as human-in-the-loop and explainable AI [10]. Systems addressing vulnerable groups have to include emotional intelligence with empathy. After all, the CPS must enable people and honor human dignity, human control, and moral accountability.

8.2.6 Environmental and Social Sustainability

The CPS has two ethical aspects: environmental and social sustainability. Whereas the CPS may happen to be environmentally conservative (e.g., smart grids or precision farming), systems, in their turn, come to be environmental problems (i.e., they are energy hogs that generate e-waste as a result of their periodic updates and sensor deployment) [11]. Environmentally, the CPS must be placed in a position of being resource-reducing and environmentally friendly. Socially, the increase in automation may lead to job displacement, especially in the construction and service sectors [12]. The CPS can be legally used to aid green technology, strategies, and labor education. Equal access to CPS technologies also has an impact on social equity. Innovation and sustainability need to be balanced to have a healthier environment and society in the long run.

8.2.7 Ethical Frameworks

Ethical frameworks provide systematic ways of designing and using the CPS in a responsible way. They help developers, policymakers, and other stakeholders strike a balance in tackling moral dilemmas of safety, privacy, responsibility, and fairness. The ethically sound design created by the IEEE is an example of transparency, well-being, and human rights [12]. The ethics guidelines by the European Commission motivate against harm, uphold human autonomy, and achieve explainability. Value-sensitive design integrates the values of stakeholders into the design process [13] and promotes the assessment of possible ethical risks performed at the earliest stage before implementation. To make the CPS useful to people and provide inclusive, people-oriented technology, these frameworks facilitate these purposes.

8.2.8 Transparency and Explainability

The domain consists of the essential ethical principles of transparency and explainability in developing and implementing the CPS, particularly in the most sensitive sectors such as healthcare, criminal justice, finance, and national security. These systems (primarily on the basis of complex artificial intelligence models) are involved in major decision-making that directly influences human lives and the outcomes of society. In such cases, the stakeholders, such as end users, developers, regulators, and the general population, need to know the reasoning and the process behind the decisions of a system. This is called explainability, and it offers confidence to the user, accountability, and good human control. Being nontransparent may result in unethical puzzles where a user can fail to contest or modify a system that may give unfair outcomes or cause harm. As an example, self-driving drones or AI clinical decision support systems could cause fatalities or even litigation in case of concealed behavior. As such, the CPS life cycle should be associated with transparency.

Some of the strategies to enhance explainability are interpretable machine learning models, e.g., decision trees, rule-based reasoning, creating user-friendly logs and dashboards, and providing natural-language explanations, which nonexperts can understand. The transparency right, or as it is commonly referred to as, the "right to explanation," is guarded by legislation including the European Union's General Data Protection Regulation (GDPR). Such law reveals that there is an ethical necessity to give clear details on automated decision processes. Transparency helps to tie the gaps between technical freedom and human responsibility, which results in behaviors being clear and comprehensible so that their actions can be regulated by ethical, legal, and social boundaries.

8.2.9 Ethical Design in Emergencies and Critical Scenarios

The ethical approach to the design of emergencies and critical scenarios is vital. CPSs applied during crisis management, critical infrastructures, or when in armed conflict must be able to make life-or-death decisions with minimal time to do so. Crises pose some moral challenges in that the urgency and magnitude of these crises intensify the ethical dilemma and complicate the decision-making process. As an illustration, a system can be requested to process patients as a priority or to close a power grid to avoid collapsing. Utilitarian-based decisions may be ethically questioned as the attempts to save a large number of people may cause the death of a few of the population. Equally, there are chances of risk magnification due to algorithmic bias, considering that in the case of system-stress misidentification or devaluation, risk will increase.

The ethical burden of determining the entity responsible for system overrides, whether it be human, automated, or remote, is considerable. One way to solve these problems is to utilize morally based logic, including so-called "ethical black boxes" that keep track of the decision. These make it possible to do postincident audits, help assign blame, and improve how the system works thereafter. In high-stakes situations, systems need to be founded on strong designs that work well and are morally sound. Ethical foresight should be a component of this.

8.2.10 Digital Divide and Access Inequality

The CPS has the potential to improve human lives; nevertheless, it also presents ethical challenges related to inequality and inclusiveness. The digital divide is one of them. It is the gap between people who have access to new technologies and people who do not. When smart-city technology changes how people move around and how services work in cities, this gap is easy to see between city and country people. This is because the countryside does not have the support it needs and is not optimized. The same thing happens to people who are better off financially. They can use contemporary CPS services such as house automation or even medical precision. On the other hand, marginalized groups cannot use these technologies because they are too expensive or the infrastructure is not good enough.

There is also an additional level, which is between people who know how to use technology and people who cannot use CPS interfaces well because they are old, uneducated, or have infirmities that make it hard for them to do so. For instance, the most advanced CPS for medical diagnostics (or robotic surgery) might only be available in the top government and private institutions, whereas the public health system might still be using old methods. This creates a two-tier care system that makes health differences worse. To encompass these ethical dimensions, the CPS must be founded on inclusivity and universal accessibility. It can be done by giving policymakers and designers the tools they need to use an equity strategy that includes providing accommodating mechanisms to underserved areas, such as subsidized finance, indigenous languages, and simple user interfaces, and cheap decentralized architectures that make it easier for more people to use. The creation of ethical CPSs must ensure that technical advancements do not exacerbate existing social imbalances; rather, the process aims to mitigate these inequalities, thereby empowering all segments of society equally.

8.2.11 Informed Consent and User Awareness

In the CPS, people often work behind the scenes without the user knowing, collecting data, observing behaviors, or making choices. It is wrong to spy on individuals when they do not realize that their actions are being watched, scrutinized, or changed by the technologies that are built into the system. In particular, these worries can be especially strong in locations such as smart homes, classrooms, hospitals, or workplaces, where the CPS will be engaging with people in small but important ways all the time. Informed consent is crucial. People have to be aware of what data are to be collected, the purposes of their usage, and the place of their storage.

In the case of the CPS, "opt-in" is preferable to "opt-out." This way, the user acknowledges the data extraction before it happens. Data consent interfaces have to be straightforward, easy to understand, and clear so that individuals of any background can choose with knowledge without having to wrangle with complicated legal or technological terms. There should also be regular reminders or notecards about data practices, especially in cases of changes or updates in policy. A well-established right to disconnect must be in place in workspaces under the control of the CPS, so that the methods of an employee are not controlled and, therefore, employees can be safe from the control of their minds, both mentally and physically. Finally, to ensure trust in the CPS, the user's perceptions should be advanced through the feeling of well-being, being respected, informed, and in control of digital communication.

8.3 Legal Implications of CPS

CPSs are taking over the modern world and have integrated themselves into every aspect of our lives. The CPS is essential for everyday objects such as sensors in self-driving cars to advanced systems such as drones and missile defense systems. The CPS is used in a variety of industries, including healthcare, automotive, defense systems, and agriculture. The employment of the CPS in the aforementioned industries is unavoidable, and the risks it poses to individuals are likewise readily apparent. Computational intelligence and physical processes are incorporated into the CPS, and this makes it susceptible to numerous forms of threats as well as challenging legal issues.

Due to the failures or the abuses of the systems involved, effects can lead as far as violations of privacy through physical injury to outright geopolitical war. The legal issues are also dependent on which domain one is considering: autonomous vehicles (AVs) have liability issues, whereas smart meters have consent and surveillance issues. The informed consent and malpractice issues are caused by the use of robotic surgery systems in healthcare. CPSs used in the industrial sectors, as in the case of the Stuxnet cyberattack, have become a driving force against international law and the rules of cyberwarfare. The issue of unauthorized data collection arises with consumer CPSs in the form of smart home devices.

These problems highlight the legal loopholes in implementing previous regulations to support intelligent and autonomous technologies. When intelligent, networked systems make judgments, it becomes hard to figure out who is responsible, what the law is, and what rights users have. To facilitate the growth of the CPS, legal rules must be amended to promote safety, equity, and openness. The subsequent instances in the real-world show how these ideas work:

- Stuxnet attack on Iran's nuclear facility
- Uber AV fatality
- Boeing 737 MAX crashes and MCAS software
- Smart meters and data privacy lawsuits
- Robotic surgery and medical malpractice
- Smart grid security breach

8.3.1 Case Studies on the Legal Dimensions of Cyberphysical Systems

(i) Stuxnet Attack on Iran's Nuclear Facility

The Stuxnet worm exemplifies a cyberphysical attack, disclosed in 2010, in which a state-sponsored outfit allegedly used it to undermine the Iranian Natanz nuclear complex. This malware targeted Siemens programmable logic controllers by manipulating speed levels on uranium-enrichment centrifuges while providing false reports to human operators. Stuxnet posed legal challenges never before seen. It is what crossed the boundary between cyber spying and war. The international-law issues that the attack brought to light were the areas of sovereignty, proportionality, and attribution, but no state actually admitted its involvement. It demonstrated the likelihood of cyber weapons to inflict real-world destruction without the need for actual intrusion. The CPS-specific cyberwarfare was not covered in any international treaty, which made the corresponding actions hard to classify and charge in the eyes of the law.

Moreover, a legal remedy could not be achieved because Iran could not take any diplomatic or legal action due to geopolitical tensions. The case identified a lapse in responsibility, software liability, and international responsibility within weaponized CPSs. International legal experts urged the creation of rules or agreements that can be enforced through the framework, such as the Tallinn Manual. Stuxnet reinforced the necessity of international collaboration to control cyberphysical weapons and the creation of sets of rules of engagement.

(ii) Uber Autonomous Vehicle Fatality

An AV of Uber struck and killed a pedestrian in Arizona in 2018, making this the first AV to kill a pedestrian. The consequences of the law highlighted these troubles of attributing responsibility within CPS settings. The backup safety driver was also charged with criminal negligence but was shown to have a software problem, including erroneous hazard classification. Uber was not found to be criminally at fault, which sparked the debate between corporate and individual proof of liability in the event of a CPS failure.

US legal systems did not provide any legislation on the use of AVs at the time, leaving loopholes in transportation safety regulation. AI-driven decision-making was put against product liability laws, and the rights of the victims were poorly discussed. This raised ethical issues such as explainability, due process, and informed deployment. Regulatory bodies, such as the National Transportation Safety Board, demanded greater control and open AV testing. The case led to pressing concerns about whether AV companies ought to bear strict liability, and it ought to be required that AVs possess audit trails. In the end, this tragedy demonstrated the incompetence of the current laws to handle autonomous actions and the division of responsibilities of the CPS.

(iii) Boeing 737 MAX Crashes and MCAS Software

The Boeing 737 MAX disasters (2018 and 2019) were caused by the malfunctioning of the MCAS (Maneuvering Characteristics Augmentation System), which is a feature of the CPS to utilize an automated process of nose management in certain flight scenarios. The cause of both accidents was traced to erroneous sensor signals misinforming the MCAS, which overrode the actions of the pilots and caused the planes to dive. These defects in the design of software, combined with the fact that the company did not warn pilots about them, caused international uproar and legal claims.

A criminal level of liability was determined, and the company was forced to pay a fine of $2.5 billion because of the lack of transparency of the CPS and the insufficient level of pilot training. The investigations revealed that Boeing's self-certification privileges had compromised the regulatory authority of the Federal Aviation Administration. The accident raised substantial legal inquiries around software validation, certification liability, and worldwide system safety assurance. Victims' families initiated lawsuits for wrongful fatalities, prompting airlines and manufacturers to reevaluate their CPS governance and risk management practices. The crashes resulted in modifications to international aircraft-certification regulations and subjected software-based CPSs to stricter scrutiny by the courts. Even a generalized regulatory framework fails to halt CPS failures when critical components have not been scrutinized carefully through all the stages of their life cycle, both legally and ethically.

(iv) The Use of Smart Meters and Lawsuits over Data Privacy

Legal battles regarding data protection and data privacy have been central over the use of smart meters installed in homes in Europe. These CPSs measure electricity consumption in real time and transmit the data to utilities to bill and analyze energy-saving profiles. There was, however, concern that the meters were recording unnecessary data, tracing users' locations, and even estimating their behaviors. The cases were filed in Germany and the Netherlands because the residents said the meters were infringing on the GDPR and their constitutional right to privacy.

The issue raised by the courts was whether collecting data without the permission of users was just, because many did not even know how much they were being surveilled. The legal cases made informed consent, restrictions in data retention, and clear processing visible. In reaction, regulators made GDPR recommendations to have minimization and anonymization of data. A guideline on CPS and Internet of Things (IoT) manufacturers was also provided by the European Data Protection Board. The case shows how good intentions of the CPS can go against the law in the event that digital rights are not sufficiently safeguarded. It highlights the overlap of data-protection legislation and the implementation of the CPS, and the need to implement regulations that balance the advancement of technologies and the rights of individuals.

(v) Medical Malpractice and Robotic Surgery

The introduction of robotized surgical systems, including the da Vinci Surgical System, is a revolutionary change in minimally invasive surgery, which has made it more precise and provided better outcomes. Nevertheless, accidents (including injuries or death) of software bugs or improper use of the surgery by the surgeon resulted in medical malpractice cases. In a number of cases in the US, patients or their relatives sued the hospitals and manufacturers of the devices because their health issues were the result of inadequate training on the system or software bugs.

These claims raised concerns about liability establishment in CPS-enabled care facilities. The question in front of the courts was what was wrong: either the surgeon (human error), the machine (technical failure), or the manufacturer (design flaw). Moreover, the issue of informed consent was brought forward, as most patients were in the dark about the fact that they were undergoing surgery with the help of a robot of CPS design. This has increased legal issues concerning proper disclosure and user knowledge. The Food and Drug Administration required postmarket surveillance and fault-reporting systems as a means of tracking the causal use and harm of the CPS. Besides, legal specialists stressed the necessity to define CPS-specific clinical trial qualifications and ethical review boards in the software update. The above cases highlight the indispensability of active legal technologies to deal with complexity in the CPS in medicine when human life is at stake due to machine decision-making.

(vi) Smart Grid Security Breach

CPS technologies are connected in smart grids to modernize power systems by automation, real-time monitoring, and intelligent management. They have, however, become targets of advanced cyberphysical threats. One of the most well-known ones happened in Ukraine in 2015, when hackers attacked the country's power grid, disrupting the power to more than 230,000 people. The malware was reported to be BlackEnergy that interfered with SCADA (supervisory control and data acquisition) systems and took away control from the operator, limiting their attack.

The first episode of a cyberattack that was able to cause the disintegration of a power grid demonstrated the vulnerability of the CPS of critical infrastructures. The case included a lot of intricate legal issues, both at home and abroad. The attack was foreign, which caused tensions between countries and brought up questions about international humanitarian law, such as the Tallinn Manual's rules for cyber warfare. Ukraine did not have a strict policy when it came to CPS-oriented cybercrime in terms of legal punishment and compensation; therefore, there was a legal obstacle at home.

The event also showed that the legal system has not kept up with the way that physical and digital infrastructure are coming together in the national infrastructure. The case started to have an effect on regulators all over the world. They talked about the prospect of making grid operators follow required compliance regimes, mandated cybersecurity assessments, and tighter vendor liability rules in case of software bugs. It also encouraged governments to demarcate legal limits to cyber sabotage and explain whether attacks in that direction can be counted as acts of war by international standards. The case is a vivid warning of the imperative to establish binding legal provisions during the elimination, stabilization, and defense of the CPS in the availability of fundamental services such as power distribution.

8.3.2 Evolving Legal Standards and Communication Protocols in CPSs

CPSs are becoming more and more interwoven into people's lives; hence, legislation about the CPS should be sound to safeguard people's privacy and system security. Legal frameworks must be designed to safeguard people's rights and obligations [14]. The rights related to digitization were first created by the Russian legislature. They allude to the legally recognized liability rights known as "digital rights." Large volumes of information are also produced by numerous inventions; for instance, newly created sensors gather and process enormous amounts of data. As a result, the ISO (International Organization for Standardization) published guidelines in 2018 for managing large amounts of data that are gathered or detected by sensors and sent or processed by the CPS.

Transmission of data is driven by protocols. Notably, the CPS relies heavily on data exchange between independent things, or the IoT. Regrettably, despite the technology's long history, clear legal and technical regulations are still lacking. The information technology regulations were drafted by Technical Committee 194.

IoT general guidelines (Russian Federation Preliminary National Standards, 2020), the CPS's data transfer model is subject to change. Internet protocols such as Internet protocol (IP), IPv4, IPv6, X.25, IP security (IPsec), and internetwork packet exchange (IPX) are the most commonly used protocols present in the Open Systems Interconnection (OSI) model at the network level.

In addition, protocols such as data distribution service (DDS), constrained application protocol (CoAP), user datagram protocol (UDP), extensible messaging and presence protocol (XMPP), and message queuing telemetry transport (MQTT) protocols are utilized for the exchange of data. In the United Kingdom, to enhance the cybersecurity of CPS users,

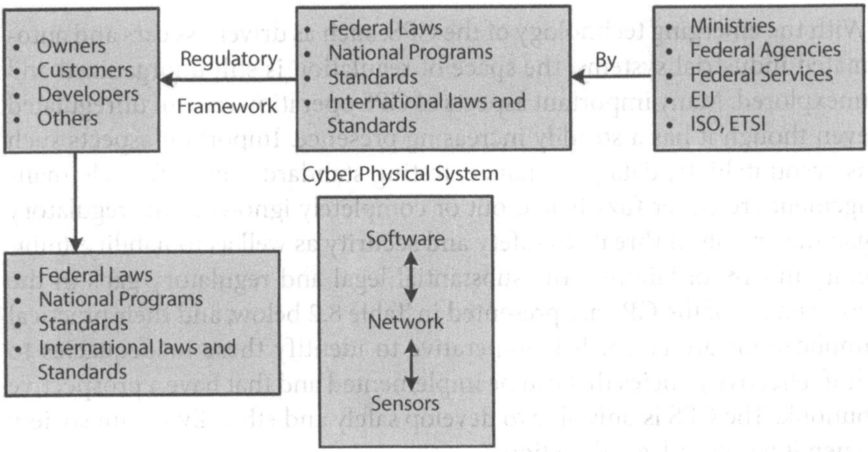

Figure 8.2 Legal implications of CPS [15].

the UK Department for Science, Innovation and Technology released its Code of Practice for Consumer IoT Security.

Two key components of data security in India are the National Cyber Security Policy 2023 and the Information Technology (IT) Act. The Computer Emergency Response Team-India (CERT-In), in accordance with the IT Act, offers instructions for keeping an eye on, identifying, preventing, and handling cybersecurity events. The goal of the 2023 National Cyber Security Policy is to instill a high level of confidence in IT systems. To protect and strengthen the country's critical information infrastructure, it also seeks to strengthen the regulatory environment. Figure 8.2 represents the structure of the legal framework in the CPS.

One of the main issues that the law must consider is the legal issues around individuals bearing responsibility for personal injuries or fatalities caused by a CPS failure. In 2018, a pedestrian died in Arizona after Uber's XC90, an AV with intelligent automated driving assistance, malfunctioned. Although the pedestrian was successfully located by the sensors in this instance, it was not deemed a system threat [16]. Additionally, the emergency braking system was not activated because of a technical glitch in the system. A few important issues that the law has to address include:

i) The emotional, financial, and physical harm brought on by a CPS's malfunction
ii) Threatening or injuring someone by abusing the authority allowed to access the interconnected distributed systems

8.3.3 Current Gaps in Laws Governing CPS

With the emerging technology of the CPS, such as driverless cars and automated industrial systems, the space of regulation is still unorganized and unexplored. Many important aspects of CPS operation remain unregulated even though it has a steadily increasing presence. Important aspects such as accountability, data governance, testing standards, and life cycle management are either fuzzily laid out or completely ignored. This regulatory gap may result in threats to safety and security as well as to liability ambiguity in case of failures. The substantial legal and regulatory gaps in the governance of the CPS are presented in Table 8.2 below, and their practical implications are given. It is imperative to identify these inadequacies to draft effective policies that can be implemented and that have a prospective outlook. The CPS is only able to develop safely and ethically within society when it has clear legal directions.

8.3.4 Evolving Legal Frameworks for Secure Cyberphysical Systems

As the world continues to expand its CPSs and integrate them into our daily lives, it is time to revamp and either implement or reinforce the legal infrastructure that governs the design, deployment, and use of these systems. Existing legislation tends to ignore the dangers of the CPS, particularly safety, accountability, privacy, and data security. Table 8.2 demonstrates the

Table 8.2 Current gaps in CPS legislation and implication.

Sl. no.	Category	Gap	Implication
1.	Accountability and liability	Indeterminate assignment of fault	Prevents legal action following accidents
2.	Security criteria	Shortage of cybersecurity rules and regulations	Increases susceptibility to attacks
3.	Real-time data	No common regulation for the data collection methodology in CPS	Lack of transparency in the choice made by the autonomous CPS
4.	Testing and validation	CPS reliability cannot be confirmed by a single, consistent policy	In rare yet crucial edge scenarios, CPS may malfunction
5.	Liability of life cycle	Typically, regulations do not address CPS upgrades, decommissioning, or aging components	Long-term effects on the environment, safety, and security are ignored
6.	Emergency overrule and manual control	Not authorized in all domains	Safe landing mechanisms are absent from CPS used in drones or autonomous vehicles

key deficiencies in the CPS legislation in place and the implications of the inadequate coverage.

With this, there is some response as some countries are initiating CPS-focused regulations. In the US, self-driving cars have been provided with elaborate regulations by the National Highway Traffic Safety Administration. The main goals include system safety, auditability, real-time monitoring, and user protection. These regulations are meant to ensure that AVs operate within regulations and to ensure that they are under persistent monitoring.

A more proactive approach has been undertaken by the European Union, which has passed the EU Artificial Intelligence Act in 2024. In this act, specific standards are placed on data management, data-stability evaluation, and continuous monitoring, especially for high-risk AI applications such as self-driving vehicles. The aim of the law is to enhance the trust of people and ensure the responsible use of AI and CPS technologies in the EU market.

In India, cybercrime and digital malfeasance have a guideline through the Bharatiya Nyaya Sanhita and the IT Act. Such laws, however, fall behind as the adoption of the CPS increases, covering such areas as smart infrastructure and self-driving cars. To deal with this, India had earlier introduced the Digital Personal Data Protection Act to focus on personal privacy and the security of data produced by the CPS. Although it is a positive move, India is yet to have full-fledged rules that cover all applications in the CPS, particularly in transportation. Regulations and guidelines that are particular to CPS-based mobility systems are presently being created. This is due to the fact that AVs are still in the early stages of development inside the country. These emerging regulations are expected to provide clearer guidelines on liability, cybersecurity, system validation, and ethical use.

To truly secure the CPS ecosystem, there is a critical need for harmonized, forward-looking legal instruments that address both the technical and ethical dimensions of these systems. Countries must work collaboratively to create globally aligned standards while considering local socio-economic contexts. This way, they may encourage new ideas while keeping people and infrastructure safe from the many hazards that come with the CPS [17].

8.4 Social Implications of Cyberphysical Systems

CPSs are transforming how societies work by giving intelligence to the physical structures of transportation, healthcare, education, agriculture, and urban planning processes so that they can run now and in the future. These systems ensure efficiency and automation, but they also have serious societal effects that need to be handled carefully. One of the biggest problems in society is that automation takes away jobs. The further advancement of intelligent robots has the potential to supplant workers in manufacturing, logistics, and customer service, resulting in unemployment and societal inequality [18]. Although technical and data-management work can be created, not everyone who loses a position as a result will be capable of making the switch or be given such a chance.

CPSs also bring up the issue of privacy and surveillance. There has been an increasing reliance on the use of smart homes, wearable gadgets, and monitoring systems functioning in a public setting, which gather and analyze personal information without clear permission or knowledge by the user of the system, thus resulting in the loss of autonomy and reliability. Additionally, the digital divide can be increased on the condition that only privileged layers of society can access and take advantage of CPS innovations, and marginalized groups will be forgotten.

Another very important issue is algorithmic bias; when trained with bias, the CPS can carry a bias into an already biased system; criminal justice has been shown as an example, but education is at stake as well. Social consequences extend to the ethical treatment of individuals, wherein machine decisions may disregard human feelings or specific circumstances. Safety is another huge issue; if the CPS does not work, it could cause accidents or injuries, especially with self-driving cars or medical equipment. Also, environmental problems are getting worse because more electronic trash and energy are being used, which also hurts people's health in a roundabout way.

The CPS can help physicians and patients get better at diagnosing and treating illnesses, but it can also make it harder for them to talk to each other. It is important to be accessible; the CPS should make it possible for people with disabilities and older people to get services without making things harder. It is necessary to teach and raise awareness of CPS technologies among the general public so that they may use them wisely and fight against deception. To be accepted in society, you need to build trust, be open, and get involved in social activities. Finally, people who work with the CPS should use legal and moral frameworks with a lot of enthusiasm to promote social well-being, justice, and inclusion. CPSs can improve society, but if they are not done ethically, they can make things worse and hurt people's dignity.

In addition to these social and environmental concerns, CPSs, particularly those integrated with artificial intelligence, are more susceptible to hacking and exploitation. Such systems may harbor sensitive information that can be exploited by a hacker to modify data for political or commercial gain. If anyone can get to information without any restrictions, they could leak, disseminate false information, and be used incorrectly with CPS resources. Devices within CPS ecosystems, such as smart televisions, refrigerators, and security cameras, can be commandeered and transformed into attack instruments, such as components of malware during a distributed denial of service attack [19]. The computing power provided by the CPS can be used by hackers to cause the disruption of businesses, vital services, and critical infrastructure.

The existence of these vulnerabilities proves that the CPS, albeit contributing to a solution of acute problems in the world, including aging, urbanization, and the sustainability of the environment, adds new security threats when the rules and protection are not sufficient. Thus, the implementation of the CPS should be guided by ethical principles, effective cybersecurity, open-access architecture, and clear legislation to make sure that these systems have a positive impact on society instead of a negative one.

With cities growing exponentially, so does the demand in terms of services, resources, infrastructure, and transportation. These challenges are solved by the CPS. For instance:

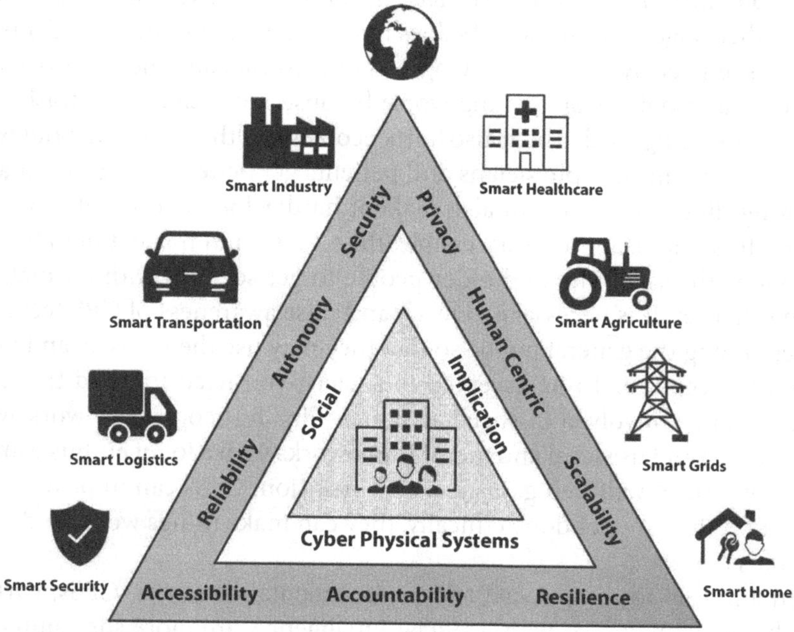

Figure 8.3 Overview of social implications of cyberphysical systems.

(i) They manage traffic with ease through real-time traffic monitoring.
(ii) They optimize electricity distribution in smart grids.
(iii) They use a variety of sensors to control waste and pollution in society.

The issue of an aging population is significantly influenced by the CPS. The need for assisted living and healthcare is growing as the number of elderly persons in society rises. Robotic assistants and wearable health monitors are two examples of how the CPS offers solutions. Furthermore, by giving real-time data on agricultural aspects including soil, weather, and crops, the CPS offers solutions to environmental sustainability issues such as pollution, climate change, and resource depletion. The transition from Society 4.0 to Society 5.0 is made easier by all of these. The CPS in Society 5.0 contributes to the development of an environment in which technology is an effective tool for improving equality, human well-being, and societal resilience by implementing technological breakthroughs with moral and social principles. A summary of the social implications of the CPS is depicted in Figure 8.3.

8.4.1 Society 4.0

Society 4.0 is the fourth phase of changes in the evolution of society, the replacement of the Industrial Revolution and the beginning of information technology. This is the period that settled deep in the late 20th and the early 21st century and was characterized by the fast pace of technological development based on the emergence of the Internet, mobile communication, cloud computing, and big data analytics. During this time, the major preoccupation was on the digitalization of industries and advocating automation, efficiency, and productivity. The implementation of the CPS, the industrial IoT, and the evaluation of smart factories increased the intelligence of manufacturing systems at the base of what became commonly referred to as "Industry 4.0." Another important trait of Society 4.0 was the lack of coupling between the cyber world (virtual) and the processes in the real world. The technologies used in the field of digital technologies worked as an additional layer to human society and industries without penetrating deeply into the everyday lives of people. Even as information systems were getting more digitalized, they were limited to such spheres as finance, telecommunications, or manufacturing and did not think about including people in the realm of applications with the benefit of society [19].

Nevertheless, as much as there was a leap in technological dimensions, Society 4.0 had a few setbacks. To begin with, digital technologies may have been centralized and complex, thus limited to those whose economies and geographical locations were disadvantaged. Urban and rural areas, in addition to developed and developing countries, had a massive digital disparity [20]. Second, the majority of systems came to the fore focusing on efficiency, rather than on human-centered design. The aspects of automation were usually without empathy and flexibility, and therefore, were hard to use by the general population. Besides, data privacy and transparency were not principal design features, resulting in trust gaps in governmental online services. The overall technological influence was not wide-ranging and all-encompassing, which contributed to increasing societal worries in regard to equity, accessibility, and human welfare. This triggered the imagination of Society 5.0, the vision leading to the balance between the growth of technology and human-based values. It seeks to build a super-smart society *via* the integration of digital, physical, and biological domains to enhance quality of life, inclusiveness, and sustainability [20].

8.4.2 Society 5.0

Society 5.0 is an innovative model in the social development era whose vision is to blend the most advanced technologies, i.e., CPS, artificial intelligence, and the IoT, into all functions of social life. Society 5.0 is markedly different in comparison with Society 4.0, the primary subject of which was industrial automation and digitalization. Society 5.0 is concerned with ecological accountability, societal fairness, and human well-being alongside technological and financial advancement. To be precise, the main goal of Society 5.0 is to develop a super-intelligent society in which cyberspace (digital world) and real space (real-world) are perfectly integrated with intelligent systems.

Under such integration, one can proactively detect human needs and provide them with personalized context-aware services. CPSs can optimize the quality of life of people and communities by using data transmitted *via* IoT stations and processed with the help of AI motifs to expand their decisions. Salient features of Society 5.0 are:

(i) **Cyberphysical integration**: The status of physical environments is sensed, processed in the cloud, and physical systems within the domains of healthcare, transportation, agriculture, etc., are either automated or facilitated.
(ii) **AI and IoT partnership**: Smart systems coordinate to predict societal needs, e.g., the provision of care to the aged or

attending to disasters, and provide specific solutions, minimizing manual effort and enhancing efficiency.

(iii) **Sustainability and resilience**: The focus is on establishing systems that not only need to be smart but naturally also environmentally conscious and capable of withstanding disruptions such as changes in the climate, pandemics, and cyberattacks.

(iv) **Inclusive innovation**: Society 5.0 advocates digital inclusion by spreading equitably the advantages of technology, narrowing the digital divide, and encouraging the involvement of all social layers and the rural population and the elderly.

(v) **Individual-centric engineering**: Innovations are crafted to enhance human dignity, autonomy, and empowerment, utilizing automation to serve humanity rather than replace it.

Society 5.0 represents a sociotechnological revolution that redefines the concept of progress. It envisions a future where individuals attain more fulfilling and significant lives through technology, addressing critical social challenges such as aging, population increase, climate change, and income disparity. It represents a transition to a new civilization where social harmony and ecological equilibrium coexist alongside economic advancement.

8.4.3 CPS: Bridging Society 4.0 and Society 5.0

The emergence of human civilization has undergone a series of significant transformations: the hunting era, the farming era, the industrial period, and presently, the highly digitized information era, sometimes referred to as Society 4.0.

This period marked a technological revolution as calculators, data processing, and communication became widespread and commodified. Nonetheless, Society 4.0 focused on technological advances, digitalization, and productivity, often neglecting human-centric principles such as accessibility, sustainability, and inclusivity. The demonstrations of the effectiveness of a more adaptive and human-oriented solution to challenges affecting the world—such as climate change, aging populations, pandemics, and economic disparities—emerged as the world continued to grow more and more demanding.

The result of this vision was Society 5.0, initially presented in Japan as a state plan and now becoming a trend in the rest of the world. Society 5.0 tries to create a so-called "super-smart society" where technology can support human well-being but not predetermine it. The foundation of such

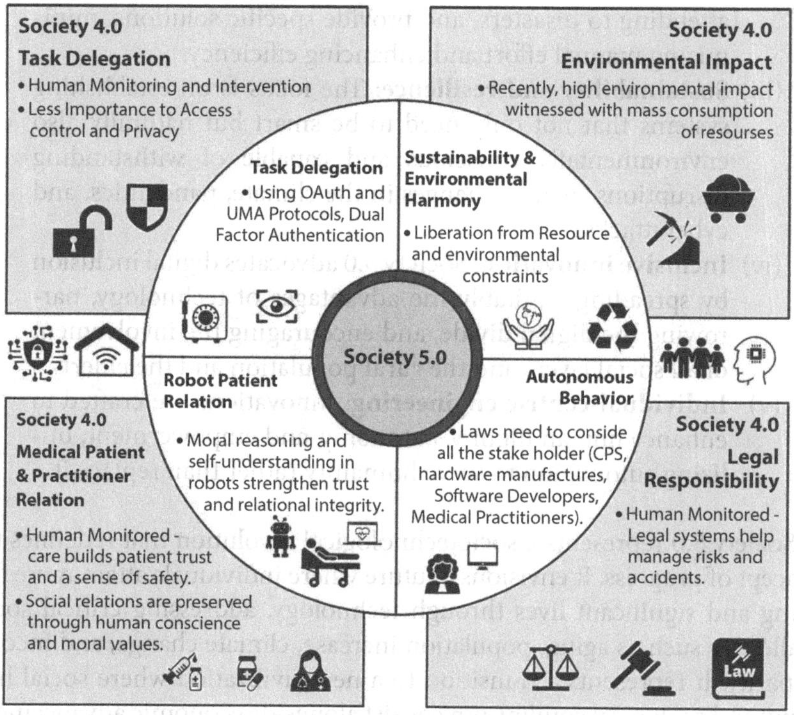

Figure 8.4 Overview of CPS implication in Society 5.0 [20].

evolution is in the CPS-intelligent, networked systems, effectively integrating the actual and digital worlds. CPSs make it possible to sense, process, and execute data in real time in different sectors, thereby supporting the personalization of services, inclusive growth, and resilient environments.

The shift of Society 4.0 to Society 5.0 is discussed along some of the most important dimensions, as shown in Figure 8.4.

In the sphere of task delegation and privacy, Society 5.0 makes use of the CPS with strong protocols such as open authorization (OAuth) and user-managed access (UMA) to increase automation and security in society compared to the weak access control of Society 4.0. Communication in the field of healthcare, where patient–practitioner relations are replaced by smart robot–patient interactions, ensures machine ethics and moral intuition to boost trust. Laws and regulations are transformed concerning their being monitored by human beings to autonomous behavior in law, inclusive of stakeholders such as CPS developers and medical practitioners.

Also, when comparing Society 4.0 and Society 5.0, it is clear that mass consumption of resources led to unacceptable environmental impact;

Table 8.3 Key differences of Society 4.0 and Society 5.0.

Sl. no.	Aspect	Society 4.0	Society 5.0
1	Focus	Efficiency and productivity	Human-centric well being
2	Role of CPS	Industrial automation	Social integration
3	Type of innovation	Novel technology	Sociotechnical
4	Challenges	Economic inequality	Human and ethical principles
5	End result	Smart manufacturing firms	Smart societies

in the latter, sustainability and environmental harmony are encouraged, which is achieved *via* streamlining the use of resources and the adoption of green technologies. Finally, the technological base of Society 5.0 consists of the CPS that will provide a shift to a future that will be not only smart and effective but also equitable, secure, and environment-friendly. Table 8.3 shows the major distinctions between Society 4.0 and Society 5.0.

8.5 Conclusion

Our modern society has developed in the context of the social, legal, and ethical consequences of the CPS. The CPS takes advantage of sociotechnical aspects informed by industry, society, and human behavior to disrupt the arbitrariness between the physical and digital worlds. In comprehending the impacts of the CPS, it is particularly important to note that the concept is part of the context of a society that is intelligent and human-centered (Society 5.0).

From the legal perspective, the CPS requires strong frameworks in terms of security, privacy, accountability, and liability. Nonetheless, most of the extant legal frameworks are configured to serve traditional systems and are, therefore, still internally insufficient to deal with more complicated circumstances arising due to CPS breakdowns or transjurisdictional data transfers. It is important to reform legal doctrines to allow for the CPS's variable nature. Because the CPS requires real-time input of data through satellites, biomedical devices, and patterns of behavior, legal systems have to clearly state regulations regarding the ownership of data, their storage, and cross-border transmission in a widely distributed network.

The CPS does promise to reduce the digital divide and improve the living status of all. Nevertheless, AI and machine learning can be used to enhance preexisting bias and yield unfair results when it comes to employment, lending, and law enforcement. These problems need to be addressed together with the cooperation of legislators, technologists, ethicists, and the population at large. Future research directions need to be based on the elaboration of flexible legislation to be updated with the emerging technologies, investing in open and fair AI in the CPS, and developing real-time ethical audit instruments. Interdisciplinary research is a necessity to develop laws that will not only positively impact technological development but also protect the health of society. By getting involved in the legal, ethical, and social aspects, the CPS can contribute to the creation of a more inclusive, ethical, and resilient Society 5.0.

AI disclosure statement: The author confirms that no artificial intelligence (AI)-generated content or AI-assisted writing tools were used in the preparation of this chapter (including summaries and the abstract). All text, analysis, and conclusions are the result of the author's original work.

References

1. Choudhury, S., Lee, M., Kurenkov, A., In Favor of Developing Ethical Best Practices in AI Research. *The Stanford AI Lab Blog*, 21, 2019.
2. Pretschner, A. and Kacianka, S., Accountability for Cyber-Physical Systems. *In Encyclopedia of Cryptography, Security and Privacy*, Berlin Heidelberg, 11, 3, 2021.
3. Klingler, C., Silva, D.S., Schuermann, C., Reis, A.A., Saxena, A., Strech, D., Ethical issues in public health surveillance: a systematic qualitative review. *BMC Public Health*, 17, 295, 2017.
4. Barhamgi, M., Perera, C., Ghedira, C., Benslimane, D., User-centric privacy engineering for the internet of things. *IEEE Cloud Comput.*, 5, 47, 2018.
5. Furrer, F.J., Safe and secure system architectures for cyber-physical systems. *Inf. Spektrum*, 46, 96, 2023.
6. Wang, Z., Xie, W., Wang, B., Tao, J., Wang, E., A survey on recent advanced research of CPS security. *Appl. Sci.*, 11, 3751, 2021.
7. Buolamwini, J. and Gebru, T., Gender shades: Intersectional accuracy disparities in commercial gender classification. *In Conference on fairness, accountability and transparency*, PMLR, 77, 91, 2018.

8. Terhörst, P., Kolf, J.N., Damer, N., Kirchbuchner, F., Kuijper, A., Post-comparison mitigation of demographic bias in face recognition using fair score normalization. *Pattern Recognit. Lett.*, 140, 332, 2020.
9. Almasoud, A.S. and Idowu, J.A., Algorithmic fairness in predictive policing. *AI Ethics*, 32, 15, 2024.
10. Vatter, J., Mayer, R., Jacobsen, H.A., The evolution of distributed systems for graph neural networks and their origin in graph processing and deep learning: A survey. *ACM Comput. Surv.*, 56, 37, 2023.
11. Pinzone, M., Albè, F., Orlandelli, D., Barletta, I., Berlin, C., Johansson, B., Taisch, M., A framework for operative and social sustainability functionalities in Human-Centric Cyber-Physical Production Systems. *Comput. Ind. Eng.*, 139, 105132, 2020.
12. Jacobs, N. and Huldtgren, A., Why value sensitive design needs ethical commitments. *Ethics Inf. Technol.*, 23, 26, 2021.
13. Uriarte-Gallastegi, N., Landeta-Manzano, B., Arana-Landín, G., Laskurain-Iturbe, I., How do technologies based on Cyber–Physical systems affect the environmental performance of products? A comparative study of manufacturers' and customers' perspectives. *Sustainability*, 14, 13437, 2022.
14. Zhernova, V.M. and Minbaleev, A.V., The integrated regulation of a cyber-physical system. *In CEUR Workshop Proceedings*, Central Europe, 16, 124, 2021.
15. Pasandideh, S., Pereira, P., Gomes, L., Cyber-physical-social systems: Taxonomy, challenges, and opportunities. *IEEE Access*, 10, 42404, 2022.
16. Tyagi, A.K. and Sreenath, N., Cyber physical systems: Analyses, challenges and possible solutions. *Internet Things Cyber-Phys. Syst.*, 1, 22, 2021.
17. Rither, A.C. and Hoxie, C.M., Legal Considerations of Cyber-Physical Systems and the Internet of Things. *Secu. and Priv. in Cyb-Physical Sys: Found., Prin and Apps.*, 9, 115, 2017.
18. Pimenta, D., Rodrigues, J.C., Oliveira, J.F., The social impact of the use of cyber-physical systems in manufacturing: An initial approach. *In Int. Work. on Ser. Orient. in Hol. and Multi-Agent Man.*, Cham: Springer International Publishing, 72, 2019.
19. Trentesaux, D. and Karnouskos, S., Engineering ethical behaviors in autonomous industrial cyber-physical human systems. *Cogn. Technol. Work*, 24, 113, 2022.
20. Patil, R.V., Ambritta, N.P., Mahalle, P.N., Dey, N., Medical cyber–physical systems in Society 5.0: Are we ready?. *IEEE Technol. Soc. Mag.*, 3, 189, 2022.

9

Addressing Challenges and Issues in Data Privacy and Cyberphysical Security Systems in Real-World Implementation

Lekshmi A. C.[1*], Koppala Guravaiah[1] and Shashidhara R.[2]

[1]*Department of Computer Science and Engineering, Indian Institute of Information Technology (IIIT), Kottayam, Kerala, India*
[2]*Samsung Research and Development Department, Bangalore, Karnataka, India*

Abstract

Cyberphysical human systems (CPHSs) are advanced systems that integrate processing, data transmission, and actuation to manage real-world processes in real time with minimal human intervention. Operating CPHS as distributed networks enables autonomous monitoring, decision-making, and automation across diverse domains, including energy, healthcare, transportation, and manufacturing. On the other hand, cyberphysical systems (CPSs) are highly vulnerable to various cyber and physical attacks due to their constrained and distributed nature. A key concern is that many CPS devices are not designed with built-in security, leaving them exposed to risks including malware, signal disruption, physical tampering, and insider attacks. Advances in areas such as federated learning, artificial intelligence–based intrusion detection, context-aware privacy mechanisms, and lightweight cryptography strengthen CPHS security. Privacy adds another layer of challenge, as conventional methods often fail to provide sufficient protection in dynamic CPHS environments. To address these issues, adaptive privacy-preserving strategies including differential privacy and user-centric mechanisms such as the self-sovereign identity concept can be deployed. The introduction of 5G and its reliable, low-latency communication capabilities are making CPHSs increasingly capable of supporting safety-critical applications, including autonomous vehicles, healthcare services, and industrial operations.

**Corresponding author*: lekshmiac.25phd21002@iiitkottayam.ac.in;
ORCID: 0009-0009-1780-2150

Anitha Velu, Prasanth Aruchamy, Raghu Ramamoorthy, Rajesh Kumar Dhanaraj and Manjula Arunraj (eds.) Cyber Physical Human Systems: Emerging Trends and Applications, (209–228) © 2026 Scrivener Publishing LLC

Keywords: Cyberphysical human system, layered CPHS architecture, data privacy, cyberphysical security systems, smart grid, intelligent transportation system, industrial Internet of Things, healthcare systems

9.1 Introduction

A cyberphysical human system (CPHS) consists of tightly and loosely coupled components that interact with both the computational and physical domains through a responsive and coordinated framework [1]. These systems play a pivotal role in driving innovations in areas including smart healthcare systems, intelligent urban systems, advanced energy grids, intelligent transportation networks, and industrial automation. The architecture of CPHSs [2] can be broadly classified into three essential layers, namely perception, transmission, and application. The modules of CPHSs are categorized into two principal classifications: sensing components and controlling components. Furthermore, various modeling approaches are applied in CPHSs, such as timed actor, event-based, lattice-based, and hybrid models to optimize the performance and flexibility of real-time operations.

Owing to the constrained and distributed characteristics of CPHSs, they are highly susceptible to diverse cyber intrusions and physical sabotage. The type of attacks in CPHSs can be broadly classified as passive attacks and active attacks [3]. As traditional privacy tools, such as encryption, may not work effectively and efficiently in dynamic CPHS environments, more adaptive privacy-preserving techniques are required [4]. User-centric privacy preservation [5], using the concept of self-sovereign identity (SSI) [6], holds significant importance in the context of CPHSs for maintaining the privacy and authentication of digital data. Because most CPHS devices do not incorporate security in their design, they contain vulnerabilities that can be exploited by attackers to launch active attacks. Here, a layered defense strategy should be implemented that includes intelligent monitoring, encryption, access controls, and even human-focused safeguards training [7]. The incorporation of state-of-the-art tools such as artificial intelligence (AI) [8], blockchain [9], and federated learning [10] makes CPHS not only smart but also secure, private, and trustworthy. This chapter delves into the challenges and issues of data privacy and cyberphysical security systems in real-world environments.

9.2 Architectural Framework of CPHS

A CPHS is a smart, interactive embedded system network that functions through the coordinated activities of its three core components: sensors, aggregators, and actuators. To manage the complexity and ensure efficient implementation, CPHS architectures are typically structured around three interconnected layers: perception, transmission, and application.

9.2.1 Perception Layer

This layer, sometimes called the recognition or sensing layer, is the lowermost layer of CPHS (Figure 9.1). It consists of various devices, including sensors, actuators, and aggregators, to acquire instantaneous data for monitoring, tracking, and analyzing the tangible world. For instance, sensors capture input parameters from the real-world locale, whereas actuators operate according to the decisions made by higher layers using the data collected by the sensors. Electrical consumption, temperature, humidity, audio, and photon signals are examples of live data collected by sensors.

To increase the security assurance level in general, end-to-end encryption is needed across all layers, which requires expensive computations and large memory resources, making it infeasible for constrained legacy devices. As a solution to this problem, lightweight security protocols and encryption methods have been introduced, which safeguard the limitations of resource-constrained devices. However, various threats such as eavesdropping, port scanning, and passive replay attacks have exploited the vulnerabilities of this layer, which undermine the confidentiality, privacy, and authentication of the system. To address these attack vectors, security measures including trust management, source authentication, and data protection can be used.

9.2.2 Transmission Layer

The intermediate layer of the CPHS architecture, commonly termed the transport or network layer (Figure 9.1). Functioning as an intermediary, this layer ensures effective exchange of data and information between the perception and application layers. Utilizing infrastructure components, namely cloud platforms, routers, Wi-Fi networks, Zigbee protocols, etc., this layer performs real-time communication and coordination among devices and services.

The security lapse in this layer increases the likelihood of various cyber threats, including denial of service (DoS), distributed denial of service (DDoS) [11], man-in-the-middle attacks, repudiation, and replay attacks, all of which jeopardize confidentiality, integrity, availability, and authentication. Suitable security measures should be implemented to successfully mitigate these attacks. For example, the vulnerability of illegitimate entry can be reduced by implementing robust password management protocols or deploying multifactor authentication mechanisms. Lightweight cryptographic techniques can be implemented for data protection and authentication in resource-constrained devices, where security and performance trade-off occurs. Additionally, secure logical networking mechanisms, such as virtual private networks, can create encrypted pathways that support data confidentiality.

9.2.3 Application Layer

This layer is the most interactive and visible layer, where data analysis, information processing, and decision-making (e.g., services such as smart waste management, autonomous vehicles, traffic control systems, infrastructure monitoring, etc.) in CPHSs take place (Figure 9.1). This layer transforms raw data into meaningful insights using computational cloud services, middleware architecture, and advanced data mining algorithms, which lead to decision-making processes. However, this layer is vulnerable to numerous sophisticated threats, such as malicious code injection, malware, trojans, worms, and buffer overflow attacks, which can severely undermine privacy, security, safety, and authentication.

Security architects should develop a comprehensive defense strategy that incorporates technologies such as intrusion detection system (IDS) and prevention system and firewalls, designed to actively detect, monitor, and block malicious traffic. The high level-layered architectural diagram of the CPHS is shown in Figure 9.1.

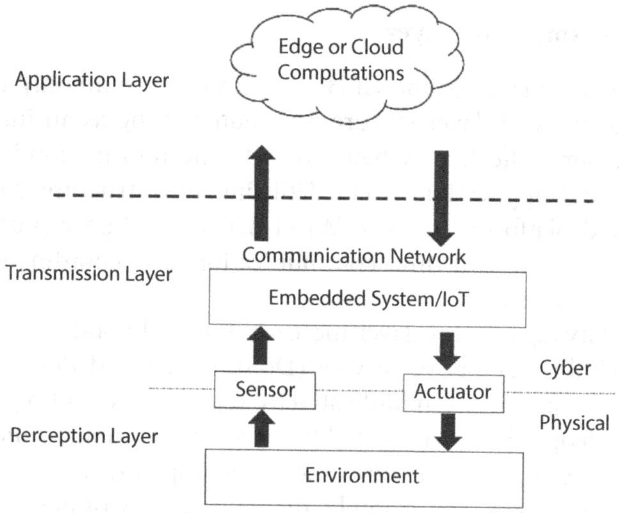

Figure 9.1 High-level layered architecture of CPHS.

9.3 Data Privacy in CPHS

CPHSs are complex systems, as they involve the interconnection of various embedded systems located at different layers of the architecture, including sensors, cloud services, and user devices, all of which have the potential to become targets for privacy threats [12]. It is very difficult to track and prevent intrusions, especially when attacks are passive in nature or do not immediately disrupt the system operations. For instance, if confidentiality of data is not protected in a wearable biosensor system that sends patient data to doctors continuously, then a hacker could intercept those data and use them for harmful purposes including injection of false information. Here, the consequences go far beyond data loss, as they directly impact people's safety and trust.

9.3.1 Data Privacy Challenges

CPHSs, including smart homes [13], medical wearables, and automated industrial machinery [14], are gathering enormous volumes of private and operational data as technology becomes more ingrained in day-to-day lives, making them attractive targets for privacy attacks. Understanding the different types of privacy attacks helps us design better protections and stay one step ahead of these threats.

i) Disclosure attack: Picture someone surreptitiously monitoring network traffic to determine who is speaking to whom rather than to see what is being said. That is essentially what a disclosure attack does. By analyzing the patterns of data flowing through a network, an attacker can guess the identities of communication endpoints. Although these attacks are hard to pull off because they require a lot of scanning and processing, they are not impossible, especially for well-equipped attackers. An effective defense strategy involves implementing robust privacy-preserving mechanisms that ensure the confidentiality of transmitted data, such that, even if an adversary successfully identifies the sender or receiver, they still cannot access the original data transmitted.

(ii) Linking attack: This attack is all about connecting the dots. It can still be dangerous even if the data are anonymized, which means that the name and other readily identifiable information are eliminated. If someone has access to two or more datasets, they might be able to piece together the identity just by comparing the information. For example, if one dataset shows the zip code and birth year and another has the medical history, a determined attacker might match them to identify the person. To defend against these attacks, smart privacy tools are needed.

(iii) Differencing attack: In this attack, rather than directly targeting the data, the attackers may attempt to circumvent regulations. A system may restrict personal queries, such as asking whether a specific individual has diabetes, as this is considered too sensitive. Rather than making a direct request, an attacker could obtain the total count of people with diabetes and then the count excluding a particular individual. By comparing the two results, the attacker can infer private information about that individual. This sneaky method is called a differencing attack, and it shows how even aggregate or summary data can leak private details. To defend against this, systems need to add a bit of randomness so the results are not exact enough for someone to reverse-engineer sensitive information.

(iv) Correlation attack: A person's data reflect relationships with family, friends, and coworkers in real life. A correlation attack takes advantage of those relationships. For instance, if an attacker has anonymized health data from several members of the same family, they might piece together someone's health status just by analyzing the overlap. These attacks do not rely on one big discovery but on subtle patterns and similarities across datasets. Preventing them means building privacy systems that understand how data points are related and limit the amount of connected information that can be exposed.

9.3.2 Privacy-Preserving Methods in CPHSs

Encryption is one of the widely adopted methods to keep data safe, where the computational infeasibility and blind randomness of cryptographic techniques are applied with the help of secret keys. Even though encryption offers strong protection, it can be resource-heavy and may not perform well when data need to be shared publicly. Another method is anonymization, where personally identifiable information is removed from a dataset before evaluating the query. This technique works efficiently for data in high-dimensional space, but it is not resilient; as the dataset becomes large and complex, there is a likelihood of reidentification of sensitive data. Different privacy-preserving methods (Figure 9.2) used in the context of CPHSs are discussed in the following paragraphs.

Differential privacy (DP) is an adaptable solution for the aforementioned problems. It is a powerful technique designed to protect individual privacy in data analysis. The idea is simple but smart: even if someone runs a query

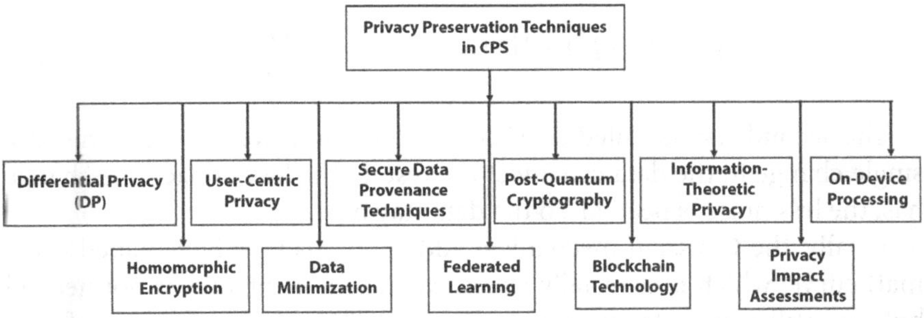

Figure 9.2 Privacy preservation techniques in CPHS.

on a dataset, they should not be able to tell whether any one person's data was included or not. This is done by introducing a calibrated amount of randomness or noise to the results so that the output looks pretty much the same whether or not a specific individual is present. In technical terms, if a randomized algorithm R, and two similar datasets D and D' that differ by only one person, then R satisfies ε -DP if:

$$\forall S \subseteq \text{Range}(R), Pr[R(D) \in S] \leq e \wedge \varepsilon \cdot Pr[R(D') \in S] \quad (9.1)$$

This simply means that the chance of getting any specific result does not change much even if one person's data are removed. That way, no outsider can confidently guess whether someone's information was used.

Over time, privacy methods have grown in two directions. First, anonymization methods such as k-anonymity and t-closeness, which try to hide identities by stripping names and similar details. But these are not foolproof. Then came DP, offering strong mathematical protection. It works in two main ways. One approach is distribution optimization, where noise is added from specific mathematical distributions such as Laplace or exponential without even looking at the data. For example, in the Laplace mechanism, the algorithm should meet privacy requirements by doing:

$$A(D) = f(D) + \text{Lap}\left(\frac{\Delta f_s}{\varepsilon}\right) \quad (9.2)$$

And for nonnumerical outputs, the exponential mechanism picks the best answer based on a utility score:

$$PrPr\left[A(D)=l\right]\propto \exp\exp\left(\frac{\varepsilon\cdot u(D,l)}{2\Delta u}\right) \quad (9.3)$$

The second way is called sensitivity calibration, tweaking how much a single change in the data can affect the result. The less impact one change has, the less noise is needed, so the data stay more useful.

Finally, the Gaussian mechanism adds noise using a bell-shaped (normal) curve, which is especially useful when perfect privacy is not needed but one still wants strong protection. If working with a function f, and know its sensitivity Δf, and want a small privacy risk δ, calculate the noise using:

$$\sigma=\frac{\Delta f\cdot\sqrt{2\log\log(1.25/\delta)}}{\varepsilon} \quad (9.4)$$

This method is significant for machine learning, where some privacy wiggle room is acceptable. What makes DP so special is how flexible it is. Whether it is for smart health devices, traffic systems, or energy networks, data should be kept intact while respecting individual privacy. With careful tuning of parameters, such as the amount of noise, data sensitivity, and required privacy level, systems can be built to be both smart and trustworthy. Even though DP efficiently solves the privacy issue, it still faces several technical issues in real-world applications.

Information-theoretic privacy is another privacy-preserving approach, based on Shannon entropy and mutual information, which provides stronger theoretical guarantees. In contrast to DP, which is based on adding noise to data, information-theoretic privacy uses entropy-based analysis, making it computationally difficult to guess sensitive data. Although DP provides strong formal guarantees and bounds on data disclosure, information-theoretic approaches give more flexibility to control how much data are revealed based on the statistical relationships between variables.

Homomorphic encryption [15] represents a significant advancement over traditional encryption methods, allowing computations to be performed directly on encrypted data without decryption, which keeps sensitive information protected throughout its lifecycle. Another approach for preserving privacy is user-centric privacy, which, based on the principle of SSI, empowers users to control their personal data. In CPHSs, this concept can be implemented through clear consent mechanisms, user-friendly data deletion tools, and transparent data-usage policies. In data minimization, privacy is protected by ensuring that only essential data are collected and processed, whereas on-device processing supports privacy by shifting computations to the source devices, reducing reliance on potentially insecure network transmissions. To preserve data integrity, secure data-provenance techniques track the origin and movement of data, making it easier to detect and trace tampering. Federated learning adds another layer of privacy by training models across decentralized nodes, thereby keeping raw data secure. Similarly, blockchain technology, owing to its decentralized and immutable characteristics, can further strengthen privacy by embedding privacy policies directly within smart contracts. To address future threats, postquantum cryptography develops techniques to counter future quantum threats and ensure long-term data security. Finally, regularly conducting privacy impact assessments helps identify vulnerabilities and strengthen overall privacy measures in CPHSs.

9.3.3 Privacy Challenges in Real-World Implementations

(i) Smart grid systems [16, 17] rely heavily on time-series data to manage and distribute electricity. This reliance raises significant privacy concerns, as energy consumption patterns can reveal individual behavior. To address this, privacy-preserving data-aggregation methods are implemented, enabling meaningful insights to be extracted from collective data without exposing personal details. Because smart grids typically require a moderate level of privacy protection, methods such as point-wise sensitivity calibration are used to safeguard individual data. Additionally, third-party aggregation systems, though cost-effective, may increase the risk of data exposure.

(ii) Transportation systems [18] rely on time-series data, collected through global positioning system (GPS) devices and sensor networks, to optimize vehicle routing and public transport operations. They use point-wise sensitivity techniques similar to smart grids but face specific privacy threats such as correlation attacks. Such attacks exploit connections across different records, such as trip patterns or location trails, to infer sensitive personal information. Because these systems operate in real time, timeliness is critical, which explains the moderate privacy requirements and reliance on third-party aggregators for rapid decision-making. Advanced privacy techniques such as trajectory perturbation and privacy-aware route planning could offer better protection while maintaining system performance.

(iii) In healthcare and medical systems [19], privacy is of paramount importance because these systems handle sensitive clinical and statistical data. These systems use a combination of instantaneous and global sensitivity calibration methods to protect patient information against inference attacks. Unlike other domains, healthcare systems cannot compromise on data accuracy, as it is critical for diagnosis and treatment. As a result, third-party data aggregation is generally not permitted, which reduces potential privacy risks. Technologies such as DP for medical records, secure federated analytics, and homomorphic encryption are being explored to maintain both privacy and clinical utility [20].

(iv) The industrial Internet of Things (IIoT) [21] relies on interconnected sensors that continuously generate time-series and statistical data, enabling extensive monitoring of critical industrial environments. Because many IIoT systems support national infrastructure, maintaining a high level of privacy is essential. IIoT systems apply both real-time and aggregate-level sensitivity calibration to protect against subtle attacks that exploit patterns over time. Due to the high priority of secure communication, these systems avoid third-party data aggregation. Promising solutions include federated learning architectures paired with DP [22], which allow edge devices to learn from local data while keeping sensitive information protected from external entities.

9.4 Cyberphysical Security Systems

A CPHS is a complex system that provides a convergence of various technologies including embedded systems, wireless sensor networks, satellite networks, Internet of Things (IoT), mobile networks, and the Internet. Advanced networking technologies, namely, 5G and beyond [23], enable ultrafast, low-latency data exchange. As CPHSs become embedded in critical frameworks such as smart grids, industry, healthcare, logistic networks, agriculture, and defense, they become prime targets for security threats. These threats may be motivated by economic, criminal, military, espionage, political, or terrorist intent. Consequently, CPHS vulnerabilities may be leveraged to launch serious threats. Ensuring the confidentiality, integrity, and availability of these systems is therefore a key security priority.

9.4.1 CPHS: Threats and Vulnerabilities

A security threat refers to any intentional or unintentional action that exploits vulnerabilities in the digital, physical, or combined infrastructure of a CPHS. Such exploitation can compromise the confidentiality, integrity, and availability of the system. The heterogeneity of CPHSs, often due to third-party integrations, increases system complexity and introduces platform-specific risks. Additionally, poor coding practices and careless universal serial bus (USB) usage can lead to serious exploits, as evidenced by the infamous Stuxnet worm. CPHS security threats can be broadly classified into cyber threats, physical threats, and cyber-physical threats (Figure 9.3).

Cyberattacks on CPHSs can originate remotely with minimal resources, making them highly dynamic and difficult to detect. This necessitates a critical evaluation of current security protocols and the adoption of strategic, multidimensional defense mechanisms. Effective cybersecurity in CPHSs must be analyzed from multiple perspectives: information-centric (protecting data during storage, transmission, and processing), function-oriented (ensuring safe incorporation of cyber and physical modules), and threat-focused (recognizing and alleviating risks to confidentiality, integrity, availability, and accountability) [24]. In contrast, physical threats in CPHSs arise from both intentional and accidental actions that compromise the physical integrity or operational reliability of system components. Finally, cyberphysical threats represent an integrated class of risks that simultaneously affect the digital and physical domains of CPHSs, making them especially challenging to detect and mitigate.

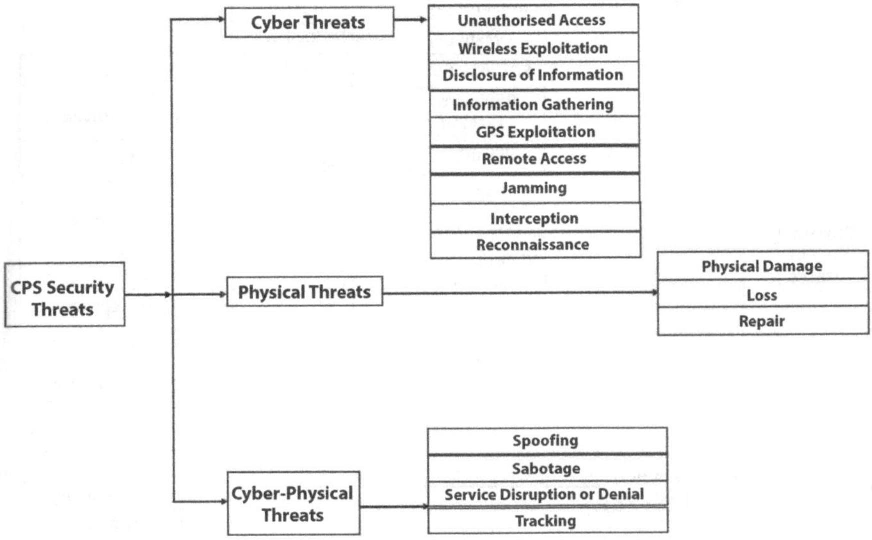

Figure 9.3 Hierarchical diagram of CPHS security threats.

Tracking threats in CPHSs highlight how attackers can compromise both privacy and operational security. By gaining physical access or attaching malicious hardware, adversaries can trace movements, extract sensitive data, or infer system behavior. Mitigating these risks requires an integrated approach to design and technology, incorporating encrypted communication, physical shielding, and behavioral anomaly detection.

9.4.2 Security Attacks in CPHS

CPHSs are increasingly targeted by a diverse range of threats that affect both the cyber and physical aspects of the architecture (Figure 9.4). Because these systems are becoming vital to critical infrastructure and industrial operations, understanding and categorizing such threats is essential for developing robust, layered defenses [25].

(i) Physical attacks on CPHS: Physical attacks on CPHSs exploit tangible system components such as hardware interfaces, facility access points, and surveillance infrastructure. Common tactics include deploying malware *via* infected USB drives, leveraging insider threats from disgruntled employees, or executing physical breaches such as wiretapping and unauthorized access through cloned keycards. Attackers may impersonate authorized personnel, manipulate CCTV (closed-circuit television)

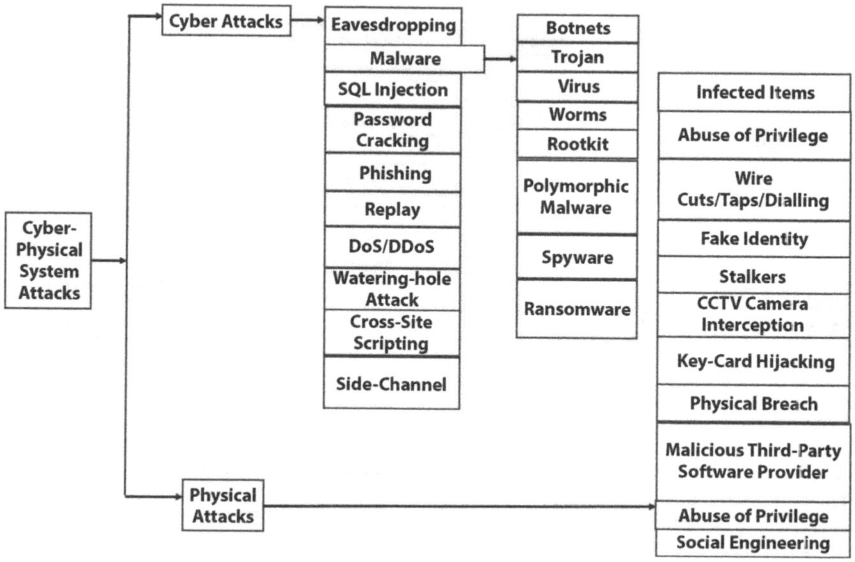

Figure 9.4 Hierarchical diagram of CPHS security attacks.

systems, or use social engineering techniques such as baiting and tailgating. These actions can result in serious aftereffects, such as data theft, operational disruptions, or complete shutdowns of critical infrastructures such as power grids or water treatment facilities. Additionally, adversaries may compromise third-party software vendors to insert malware into legitimate industrial control systems (ICSs) updates, demonstrating how physical attacks increasingly intersect with sophisticated cyber infiltration.

(ii) Cyberattacks on CPHS: Cyber threats against CPHSs exploit vulnerabilities in software, networks, and human factors, including eavesdropping, injecting malicious code through SQL injection or cross-site scripting, password cracking (brute force, rainbow tables), and phishing campaigns targeting critical personnel. More sophisticated attacks include hijacking a system's control flow, injecting fake data, or overwhelming systems with DoS or DDoS attacks using botnets [26]. Malware such as trojans, rootkits, and ransomware can seriously disrupt operations, steal confidential data, or extort organizations by encrypting critical system files. Advanced forms of threats are even harder to detect, namely polymorphic malware that constantly changes its form or spyware that hides in systems for long periods. Side-channel attacks go a step further, leveraging physical signals such as power consumption or electromagnetic leaks to extract data. These threats underline the need for adaptive cybersecurity strategies in CPHSs, combining real-time threat detection, encryption, resilient coding methodologies, and continuous threat monitoring to ensure system resilience against evolving cyber adversaries.

9.4.3 Security Challenges in Real-World Implementations

CPHSs encounter a broad spectrum of security challenges in real-world implementation, as highlighted in various studies. These challenges vary across different CPHS domains, reflecting the complexity and diversity of the applications. Addressing them requires a multidisciplinary approach that blends system modeling, practical testing, and flexible security strategies.

(i) The smart grid, comprising power generation, transmission, and distribution, represents one of the most vital and complex areas within the CPHS domain. Its real-time operational demands make it particularly vulnerable to cyber threats. Attacks targeting various segments of the grid, such as the Aurora attack on generation systems, state-estimation attacks on transmission lines, or privacy breaches in smart meters, can result in large-scale cascading failures. Defensive strategies in this context typically focus on three pillars: protection (e.g., encrypted communications), attack detection (e.g., IDSs), and mitigation (strategies to minimize the impact of an attack). Researchers have adopted a layered perspective on grid security that includes device-level safeguards, network integrity measures, and secure dispatch operations. Securing the entire lifecycle of grid data, from generation by devices such as phasor measurement units and smart meters to their storage and processing, is crucial. To this end, tools such as data analytics, visualization, statistical modeling, and data mining are used to detect anomalies and extract actionable insights that enhance grid resilience [27].

(ii) Embedded medical devices such as pacemakers and drug delivery systems pose unique security and privacy challenges due to their safety-critical nature. Threats target telemetry interfaces and can involve both passive (eavesdropping) and active (data-injection) attacks. Defense strategies involve biometric authentication, wearable security devices, and pattern-based anomaly detection. However, device constraints such as power limitations and emergency accessibility complicate the implementation of robust cryptographic protections [28].

(iii) Industrial Control Systems (ICSs) [29] control industrial and manufacturing processes and differ significantly from IT networks in terms of architecture and vulnerabilities. Protocol-related flaws in Modbus, distributed network protocol, and IEC 61850, among others, have been widely studied. Researchers have proposed IDS based on packet analysis, system state monitoring, and SCADA (supervisory control and data acquisition) logs. Additionally, new control architectures focus on runtime verification, trusted computing bases, and timing models to detect unsafe operational states. Physics-based detection methods leverage the inherent physical properties of ICSs for anomaly detection. Effective implementation requires a system model, trust assumptions, statistical tests, and metrics for detection quality. Researchers have studied such methods in domains such as electricity theft, automated vehicles, and industrial automation. Despite the promise, challenges include testbed limitations, assumption variability, and difficulty in sensor placement for optimal detection performance. Security in IIoT emphasizes integrity checking for both software and hardware. Remote management capabilities increase the risk of exposing sensitive data to external services, making local management strategies and privacy-preserving techniques essential. Lightweight but robust mechanisms are needed to maintain data integrity without compromising system performance.

(iv) Intelligent transportation systems [30] integrate sensing, communication, and automation technologies but introduce challenges in misbehavior detection. Effective detection relies on linkability between on-board units and messages to identify inconsistencies. Vehicular *ad hoc* networks form the communication backbone for vehicle-to-vehicle (V2V) and vehicle-to-infrastructure interaction. Researchers propose model-based security strategies using Markov chains, trust models, and mechanical analysis. The focus is on preventing attacks that disrupt vehicular coordination and the reliability of communication links. The agile and decentralized nature of transportation systems makes them difficult to secure. Issues such as real-time constraints, nonstandardized protocols, and scalability complicate security design. Infotainment systems and in-vehicle networks present additional vulnerabilities that require integrated security frameworks [31]. Unmanned aerial systems and modern aircraft are vulnerable to both cyber and physical threats. Attacks on communication protocols such as automatic dependent surveillance broadcast and Internet protocol over asynchronous transfer mode (IP-ATM) can lead to severe disruptions. Researchers advocate for authentication, location verification, and RFF (radiofrequency fingerprint) to enhance aviation cybersecurity. Drone-specific challenges [32] include GPS spoofing, command injection, and physical theft, requiring lightweight encryption and intrusion detection mechanisms.

(v) Smart homes enhance user interaction with the grid by enabling dynamic energy management and supporting demand response programs. Appliances communicate with smart meters and grid operators, raising new security risks. Security goals such as confidentiality, integrity, and availability guide the identification of active (e.g., tampering) and passive (e.g., eavesdropping) attacks. Common defenses include encryption, digital watermarking, and intrusion detection, although research on effective response strategies postdetection remains limited.

9.5 CPHS Risk Evaluation

CPHS risk evaluation is a critical process that helps ensure the safety and reliability of a CPHS by carefully identifying and managing risks. It begins with risk management, where potential threats are recognized, analyzed, prioritized, and monitored to keep systems secure and resilient over time. Risk assessment then determines the potential severity of each threat, classifying them as high, medium, or low impact to guide appropriate responses. Both quantitative and qualitative analysis methods are applied to understand the nature and scale of these risks, supporting informed decision-making.

9.6 Future Directions and Recommendations

In energy systems, especially smart grids [33], balancing privacy and accuracy remains a significant hurdle. Other use cases, such as the auction of microgrid resources and secure firmware updates, require privacy-preserving techniques to protect user identities and proprietary data. In resource-constrained microgrids, lightweight DP mechanisms are essential to ensure privacy while minimizing operational costs, as traditional methods are unsuitable for these environments. In smart transportation systems, privacy preservation is critical in applications such as live traffic reporting and V2V communication [34]. Although real-time location data enable better route planning, they also risk exposing users to location-tracking threats. Adaptive DP offers promising solutions by perturbing data before they are shared with route-planning applications. Similarly, V2V communication, though enhanced by modern wireless technologies, faces vulnerabilities due to encryption limitations. Healthcare systems are another area where CPHSs benefit from privacy. Body sensor networks collect sensitive health data in real time, and using encryption alone is often too resource-intensive for low-memory devices. But challenges remain in deploying noise-adding privacy mechanisms on constrained hardware, warranting further research in this domain. In IIoT, auction systems for industrial big data are vulnerable to insider threats, and privacy mechanisms offer robust mathematical protection. Similarly, in big data environments, defining intuitive privacy models, optimizing composition theorems, and dealing with high-dimensional data are open challenges. Emerging DP paradigms such as local, concentrated, and Bayesian DP offer new directions. In AI and cloud platforms, merging DP with machine learning models or scalable cloud systems can enhance privacy without incurring significant computational overhead.

The security requirements of new 5G services and industry-specific deployments, including smart cities, autonomous vehicles, healthcare, and Industry 4.0, demand low latency, high reliability, and advanced protection mechanisms. Security in this context must address threats across the physical, transmission, and application layers. In addition to these requirements, focus must also be placed on the technological advancements brought by

5G and beyond, including multiaccess edge computing, network function virtualization, software-defined networking, and network slicing.

9.7 Conclusion

A CPHS represents a critical incorporation of cyber and physical systems. The aforementioned system offers various domains of applications, including smart grids, healthcare, transportation, and industrial automation. Due to the distributed and resource-constrained nature of CPHSs, these systems are vulnerable to a wide spectrum of cyber and physical threats, demanding robust and adaptive security strategies. Contemporary approaches such as context-aware DP, federated learning, AI-driven intrusion detection, and lightweight cryptography are increasingly being explored and applied within CPHS environments. The adoption of 5G, particularly its ultrareliable low-latency communications, provides significant benefits for time-sensitive domains such as autonomous vehicles and industrial control. In conclusion, CPHS environments are heterogeneous, resource-constrained, and often safety-critical, so general-purpose solutions must be adapted.

AI disclosure statement: The author confirms that no artificial intelligence (AI)-generated content or AI-assisted writing tools were used in the preparation of this chapter (including summaries and the abstract). All text, analysis, and conclusions are the result of the author's original work.

References

1. Yaacoub, J.P.A., Salman, O., Noura, H.N., Kaaniche, N., Chehab, A., Malli, M., Cyber-physical systems security: Limitations, issues and future trends. *Microprocess. Microsyst.*, 77, 103201, 2020.
2. Yu, Z., Kaplan, Z., Yan, Q., Zhang, N., Security and privacy in the emerging cyber-physical world: A survey. *IEEE Commun. Surv. Tutor.*, 23, 1879, 2021.
3. Petroulakis, N.E., Askoxylakis, I.G., Traganitis, A., Spanoudakis, G., A privacy-level model of user-centric cyber-physical systems, in: *Int. Conf. on Human Aspects of Info. Sec. Pri. and Trust*, Berlin, 8, 338, 2013.
4. Hassan, M.U., Rehmani, M.H., Chen, J., Differential privacy techniques for cyber physical systems: A survey. *IEEE Commun. Surv. Tutor.*, 22, 746, 2019.
5. Yogi, M.K. and Chakravarthy, A.S.N., A novel user centric privacy mechanism in cyber physical system. *Comput. Secur.*, 149, 104163, 2015.

6. Preukschat, A. and Reed, D., *Self-sovereign identity*, pp. 12–504, Manning Publications, United States of America, 2021. ISBN: 9781617296598.
7. Habibzadeh, H., Nussbaum, B.H., Anjomshoa, F., Kantarci, B., Soyata, T., A survey on cybersecurity, data privacy, and policy issues in cyber-physical system deployments in smart cities. *Sustain. Cities. Soc.*, 50, 101660, 2019.
8. Androutsopoulou, M., Carayannis, E.G., Askounis, D., Zotas, N., Towards AI-Enabled Cyber-Physical Infrastructures, Challenges, Opportunities, and Implications for a Data-Driven eGovernment Theory, Policy, and Practice. *J. Knowl. Econ.*, 24, 38, 2015.
9. Al-Ghuraybi, H.A., AlZain, M.A., Soh, B., Exploring the integration of blockchain technology, physical unclonable function, and machine learning for authentication in cyber-physical systems. *Multimed. Tools Appl.*, 83, 35629, 2024.
10. Quan, M.K., Pathirana, P.N., Wijayasundara, M., Setunge, S., Nguyen, D.C., Brinton, C.G., Poor, H.V., Federated learning for cyber physical systems: a comprehensive survey. *IEEE Commun. Surv. Tutor.*, 42, 2457, 2025.
11. De Donno, M., Dragoni, N., Giaretta, A., Spognardi, A., DDoS-capable IoT malwares: comparative analysis and Mirai investigation. *Secur. Commun. Netw.*, 1, 7178164, 2018.
12. Saheed, Y.K. and Misra, S., CPS-IoT-PPDNN: A new explainable privacy preserving DNN for resilient anomaly detection in Cyber-Physical Systems-enabled IoT networks. *Chaos Solitons Fract.*, 191, 115939, 2025.
13. Sookhak, M., Tang, H., He, Y., Yu, F.R., Security and privacy of smart cities: a survey, research issues and challenges. *IEEE Commun. Surv. Tutor.*, 21, 1718, 2018.
14. Zheng, K., Zheng, Q., Chatzimisios, P., Xiang, W., Zhou, Y., Heterogeneous vehicular networking: A survey on architecture, challenges, and solutions. *IEEE Commun. Surv. Tutor.*, 17, 2377, 2015.
15. Sharma, M., Enhancing security and privacy in cyber-physical systems: challenges and solutions, in: *IEEE Ann. Comp. and Comm. Workshop and Conf.*, 14, 0682, 2024.
16. Ghosh, S., Maiti, S., Mukhopadhyay, D., Dey, S., Differentially Private Real-Time Pricing Control for Smart Grids. *ACM Trans. Cyber-Phys. Syst.*, 9, 30, 2025.
17. Liu, J., Xiao, Y., Li, S., Liang, W., Chen, C.P., Cyber security and privacy issues in smart grids. *IEEE Commun. Surv. Tutor.*, 14, 981, 2012.
18. Alam, T., Data privacy and security in autonomous connected vehicles in smart city environment. *Big Data Cogn. Comput.*, 9, 95, 2024.
19. Rushanan, M., Rubin, A.D., Kune, D.F., Swanson, C.M., Sok: Security and privacy in implantable medical devices and body area networks, in: *IEEE Symp. on Sec. and Pri.*, 12, 524, 2014.
20. Al-Ghuraybi, H.A., AlZain, M.A., Soh, B., Ensuring authentication in Medical Cyber-Physical Systems: A comprehensive literature review of

blockchain technology integration with machine learning. *Multimed. Tools Appl.*, 83, 35673, 2024.

21. Wolf, M. and Serpanos, D., Safety and security in cyber-physical systems and internet-of-things systems. *Proc. IEEE*, 106, 20, 2017.
22. Jia, B., Zhang, X., Liu, J., Zhang, Y., Huang, K., Liang, Y., Blockchain-enabled federated learning data protection aggregation scheme with differential privacy and homomorphic encryption in IIoT. *IEEE Trans. Ind. Inform.*, 18, 4049, 2021.
23. Rani, S., Kataria, A., Kumar, S., Karar, V., A new generation cyber-physical system: A comprehensive review from security perspective. *Comput. Secur.*, 148, 104095, 2025.
24. Humayed, A., Lin, J., Li, F., Luo, B., Cyber-physical systems security-A survey. *IEEE Internet Things J.*, 4, 1802, 2017.
25. Ashibani, Y. and Mahmoud, Q.H., Cyber physical systems security: Analysis, challenges and solutions. *Comput. Secur.*, 68, 81, 2017.
26. Kolias, C., Kambourakis, G., Stavrou, A., Voas, J., DDoS in the IoT: Mirai and other botnets. *Comp.*, 50, 80, 2017.
27. Alguliyev, R., Imamverdiyev, Y., Sukhostat, L., Cyber-physical systems and their security issues. *Comput. Ind.*, 100, 212, 2018.
28. Vinayasree, P. and Reddy, A.M., A reliable and secure permissioned blockchain-assisted data transfer mechanism in healthcare-based cyber-physical systems. *Concurr. Comput. Pract. Exp.*, 37, e8378, 2025.
29. Xu, H., Yu, W., Griffith, D., Golmie, N., A survey on industrial Internet of Things: A cyber-physical systems perspective. *IEEE Access*, 6, 78238, 2018.
30. Koscher, K., Czeskis, A., Roesner, F., Patel, S., Kohno, T., Checkoway, S., Experimental security analysis of a modern automobile, in: *IEEE Symp. on Sec. and Priv.*, 3, 447, 2010.
31. Woo, S., Jo, H.J., Lee, D.H., A practical wireless attack on the connected car and security protocol for in-vehicle CAN. *IEEE Trans. Intell. Transp. Syst.*, 16, 993, 2014.
32. Davidson, D., Wu, H., Jellinek, R., Singh, V., Ristenpart, T., Controlling {UAVs} with sensor input spoofing attacks, in: *USENIX Work. on Off. Tech.*, 16, 234, 2016.
33. Rahman, M.H., Wuest, T., Shafae, M., Manufacturing cybersecurity threat attributes and countermeasures: Review, meta-taxonomy, and use cases of cyberattack taxonomies. *J. Manuf. Syst.*, 68, 196, 2023.
34. Haddaji, A., Ayed, S., Fourati, L.C., IoV security and privacy survey: issues, countermeasures, and challenges. *J. Supercomput.*, 80, 23018, 2024.

10

Mist and Edge Computing Cyberphysical Human-Centered Systems for Industry 5.0 Application

Prajwal B.[1], Raghu Ramamoorthy[1*], Anitha Velu[2] and P. Bindhu Madhavi[3]

[1]*Department of Computer Science and Engineering, The Oxford College of Engineering, Bengaluru, Karnataka, India*
[2]*Department of Electronics and Communication Engineering, Sri Sairam College of Engineering, Bengaluru, Karnataka, India*
[3]*Department of Computer Science and Engineering, T. John Institute of Technology, Bengaluru, Karnataka, India*

Abstract

Industry 5.0 refers to an extension of Industry 4.0, indicating a shift from full automation toward human–machine collaboration, utilizing a set of intelligent systems that enrich overall human abilities rather than replacing them. This chapter focuses on how computing terminologies such as mist computing (MC) and edge computing (EC) are applied in cyberphysical human-centered systems to provide real-time and decentralized decision-making. In the perspective of Industry 5.0, incorporating EC and MC with CPHSs creates truly responsive and intelligent environments. Intelligent environments allow for real-time collaboration between humans, machines, and smart systems. EC and MC are similar; EC comprises processing done at the edge, meaning closer to the devices. MC goes even further, embedding intelligence directly into Internet of Things nodes and microcontrollers. These smart systems reduce overall latency and bandwidth requirements while improving the responsiveness of the entire system, making them more suitable for adaptive, environment-based industrial applications. Notably, this chapter addresses some key Industry 5.0 applications, namely, the manufacturing sector, healthcare, finance, shipping, and farming. Moreover, the chapter

**Corresponding author*: raghuace85@gmail.com; ORCID: 0000-0002-6378-0724

Anitha Velu, Prasanth Aruchamy, Raghu Ramamoorthy, Rajesh Kumar Dhanaraj and Manjula Arunraj (eds.) Cyber Physical Human Systems: Emerging Trends and Applications, (229–248) © 2026 Scrivener Publishing LLC

presents challenges currently being addressed in Industry 5.0 using solutions such as federated learning, secure edge artificial intelligence, and privacy-preserving computation.

Keywords: Mist computing, edge computing, CPHCS, Industry 5.0, IoT

10.1 Introduction

The perception of Industry 4.0 was first hosted by the German administration in 2011 [1]. Industry 4.0, also recognized as the fourth industrial revolution, has led to a massive change in the industrial landscape. This change has been primarily due to the amalgamation of cyberphysical systems (CPSs), cloud computing (CC), Internet of Things (IoT), and artificial intelligence (AI) into traditional construction and assembly environments. Industry 4.0 mainly emphasizes automation, interconnectivity between entities, machine knowledge, and real-time data, which can self-optimize based on scenarios to transform a factory into a smart factory and also help in predictive management. Smart factories rely on embedded sensors, actuators, and a large number of processors to collect very large aggregates of data in practice. These data gathered from the sensors are utilized in advanced machine learning (ML) algorithms for data-driven decision-making without any human intervention. The following describes a set of key technologies that are used or considered to be a major part of Industry 4.0.

- **CPS**: CPS denotes a type of system or computational system where there is integration between physical processes as well as computation [2].
- **IoT and IIoT**: IoT and IIoT (industrial IoT) play a huge role in enabling communication between machines, systems, and humans.
- **CC**: Industry 4.0 uses CC, which is like a central server or data store for additional processing and data analytics.
- **Big data and analytics**: As discussed above, a very large volume of data is created due to IoT sensors and actuators, which lead to big data and analytics, which help in deriving insights from prompt and original data and also help in understanding the data, which can be used for decision-making.

Some of the other key technologies are ML and autonomous machines, i.e., machines that require minimal human input and interaction to do the work, etc. The set of applications presented by Industry 4.0 is mentioned below:

- **Predictive maintenance**: Industry 4.0 helps in analytical maintenance using ML along with generated sensor data. This generally helps in managing manufacturing or processing capacity in a smart factory.
- **Smart supply chains**: Another application of Industry 4.0 is smart supply chains. This ensures that products can be delivered through blockchain and IoT. This increases transparency as well as security by providing real-time tracking of goods.
- **Automated quality control**: Processes such as product inspection can now be automated using AI and computer vision. This increases efficiency and inspection work time because machines, unlike humans, do not need to rest.

Some other popular applications of Industry 4.0 are digital twins, which attempt to simulate factory behavior in real time. Cybersecurity is another application of Industry 4.0, as security measures can be used to protect data and prevent cyberattacks because all devices are connected. Despite the various benefits and applications of Industry 4.0, the main restriction is that it focuses on replacing humans and their roles in industry, which has led to many anxieties around jobs, the ethical implications of AI, and the lack of human-based or human-centered designs. Moreover, the centralized, cloud-based processing models used in Industry 4.0 have various issues that need to be addressed, such as latency and bandwidth limitations, and applications are also a big problem when dealing with private data, as data-sensitive domains such as healthcare and finance cannot fully trust the cloud.

10.1.1 Background of Industry 5.0

To overcome the problems and contests brought by Industry 4.0, Industry 5.0 has come into the picture. Here, aspects such as human creativity, empathy, and context awareness are introduced into automated systems. Industry 5.0 is generally based on collaboration among humans and intelligent systems [3]. In the perspective of Industry 5.0, the main emphasis is on augmenting machines rather than replacing humans, as in Industry 4.0.

Table 10.1 Transformations among Industry 4.0 and Industry 5.0.

Feature	**Industry 4.0**	**Industry 5.0**
Primary concern	Automation and efficiency	Human–AI coordination
Human role	Minimal, often replaced	Central, empowered
Technology use	CPS, cloud, IoT	Edge, mist, cobots, XAI
Decision-making	Machine-driven	Human-in-the-loop
Scalability	Global *via* cloud	Localized, real-time systems
Data processing	Centralized	Decentralized (edge/mist)
Ethical AI consideration	Limited	Strong emphasis
Sustainability	Secondary concern	Core objective

Here, humans are the most vital element in the context of innovation, as they can use some cutting-edge skills, such as edge computing (EC), federated learning (FL), mist computing (MC), explainable AI (XAI), and collaborative robots (cobots). Table 10.1 explains in detail the transformations between Industry 4.0 and Industry 5.0. The most essential aspects of Industry 5.0 are described below:

(1) **Human-centricity**: Industry 5.0 aims to create machines, in addition to smart solutions for industry, that are human-centric by design and operation due to the collaboration between machines and humans. The design is human-centric by nature, meaning that humans have control over the entire process and therefore make the industry environment more reliable and efficient compared to Industry 4.0.

(2) **Resilience**: Industry 5.0 aims to build a set of robust systems that can adapt to any situation and any disruptions, such as pandemics, political tensions, or supply chain shocks. This makes the industry more resilient to the outside world and generally increases efficiency.

(3) **Sustainability**: Industry 5.0 technologies can be designed in an eco-friendly and environmentally conscious manner. This ensures that there is no harm to the environment and also helps in reducing energy and metal waste.

Figure 10.1 depicts the chief architecture of Industry 4.0 and 5.0. It explores the technology dependency of Industry 4.0 and 5.0. Furthermore, the exploration of technology and transformation in relation to Industry 4.0 and 5.0 is also incorporated in the same figure.

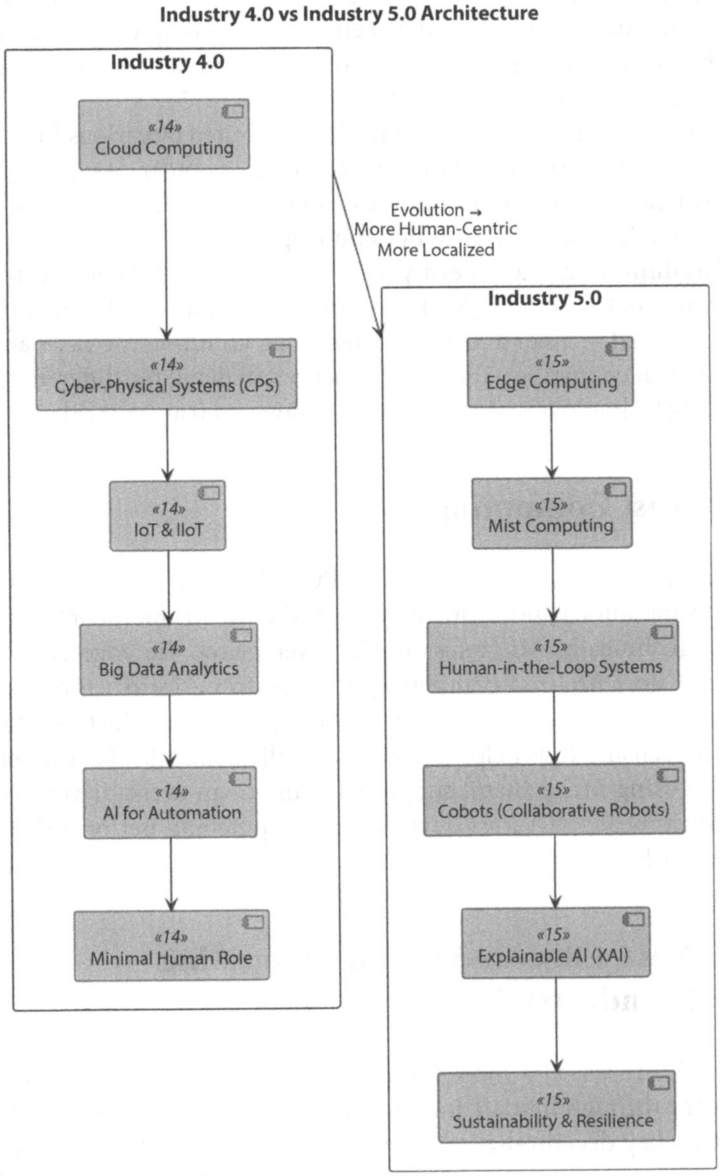

Figure 10.1 Architecture of Industry 4.0 and Industry 5.0.

10.2 Edge Computing

EC refers to a category of dispersed computing where computation occurs closer to the information source, which helps improve overall response time and also saves bandwidth. If a computation cannot be performed on the edge device, it can be offloaded to the cloud. The main advantage here is that data processing in data centers is not completely reliable, as in cloud-based computing, but data can be sent or offloaded to cloud devices when they cannot be processed on edge devices [4, 5].

EC is crucial in managing real-time data generated by various IoT devices, typically in time-critical environments. Some popular application areas include manufacturing floors, healthcare monitoring, and autonomous vehicles. EC enables real-time analytics and helps improve overall data privacy as well as diminishes the necessity for frequent and high-latency communication with central servers [6]. The following presents the different types of EC in a detailed manner and also outlines their common use cases and benefits in a detailed manner. Table 10.2 describes in detail the different types of computing technologies associated with Industry 4.0 and 5.0 [8].

10.3 Mist Computing

MC can be defined as an addition of EC, where execution is performed nearer to the information source through the use of microcontrollers and sensors. Technically, MC refers to the layer below the edge layer in the context of decentralized computing. It helps to perform various specific computations and processing closer to the edge based on the type of microsensor chosen and thus helps to reduce overall latency [9, 10, 16]. The chief benefit of using MC is that it supports the use of an event-driven architecture, which means that mist devices can compute and perform their tasks autonomously without having to wait for higher-level edge or cloud devices.

10.4 Mist, Edge, and Cloud Computing in Industry 5.0

Industry 5.0 relies on a more human-centric and collaborative approach to industry through human and machine coordination. In addition to EC and MC, a key decentralized learning approach called FL can be used in Industry 5.0, which provides local intelligence, reduces latency, and even makes context-based decisions and trains ML models without sharing any

Table 10.2 Computing technologies supported for Industry 4.0 and Industry 5.0.

Type	Definition	Architectural position	Key features	Common use cases	Advantages
Edge computing [5]	Decentralized computing is achieved near the edge of the grid and the data source	Between clouds and gadgets	Local processing reduced latency	Smart homes, real-time analytics, autonomous vehicles	Reduced latency, and bandwidth savings
Cloudlet computing [6]	Small-scale data centers at network edge	Between edge and cloud	Supports resource-intensive apps	AR/VR, gaming, mobile computing	Proximity to a user and reduced latency
Fog computing [7]	The cover of nodes among the edge and cloud that perform storage, processing, and networking	Intermediate nodes (routers, gateways)	Hierarchical, supports mobility	Industrial IoT, traffic management	Enhanced scalability and resilience
Mist computing [9, 10]	Ultraedge computing where processing occurs on the devices themselves (sensors, actuators)	Very close to or on end devices	Extremely low latency, constrained resources	Smart sensors, wearables	Real-time response, minimal communication

(*Continued*)

Table 10.2 Computing technologies supported for Industry 4.0 and Industry 5.0. *(Continued)*

Type	Definition	Architectural position	Key features	Common use cases	Advantages
Mobile edge computing (MEC) [11]	Extension of cloud computing to mobile base stations	At cellular base stations	5G integration, network- aware apps	Video streaming, autonomous cars	Low latency and location awareness
Micro data centers [12]	Small modular data centers located close to end-users	On-premise or near-user sites	Local storage, edge analytics	Retail analytics, branch offices	Privacy, security, cost reduction
Sensor- level computing [13]	Processing at the sensing unit itself	Embedded in the sensor	Embedded ML, prefiltering	Agriculture, smart cities	Energy-efficient, lightweight
Serverless edge (function-as-a-service at edge) [14]	Stateless execution of functions on demand at the edge	Edge devices or nodes	Event-driven, scalable	IoT events, image recognition	Cost efficiency, autoscaling
AI at the edge [15]	Running AI/ML models directly on edge devices	Edge nodes or embedded devices	On-device inference, low power	Surveillance, NLP, vision systems	Privacy, speed, offline capabilities

data with the cloud or fog server [17]. One of the main aspects of Industry 5.0 is the capability to instantly respond to even the smallest change in any application, be it a machine interface or a human–machine interface. EC typically uses this aspect by placing compute power close to industrial assets such as controllers or gateways, which helps reduce overall latency and also improves response times. MC can be used to provide faster computation and reduce overall latency compared to EC by using sensors and actuators to compute data locally in the mist layer without sending data to the edge devices themselves.

Industry 5.0 also deals with the concept of personalized production, which means it allows for the manufacturing of products to be heavily customized on a large scale. This is where MC comes in very handy, because these devices can adjust to the user's unique preferences if properly tuned and provided with the right hardware. Some examples are 3D printers, CNC machines, and AR-guided assembly stations, which can be equipped with MC nodes and therefore also help fine-tune their operations based on user feedback, authentication mechanisms such as biometric or facial recognition systems, or tuning based on the environment. The edge layer can use these data to provide insights to other edge devices or a cloud server or can optimize them to provide better quality and efficiency [18].

Sustainability is another key aspect in the context of Industry 5.0. In the context of EC as well as MC, because data processing is done locally, this leads to a decrease in overall bandwidth intake and generally brings out a drop in overall resource utilization. Instead of transferring all available data to a cloud server, MC can be used, which helps in lightweight

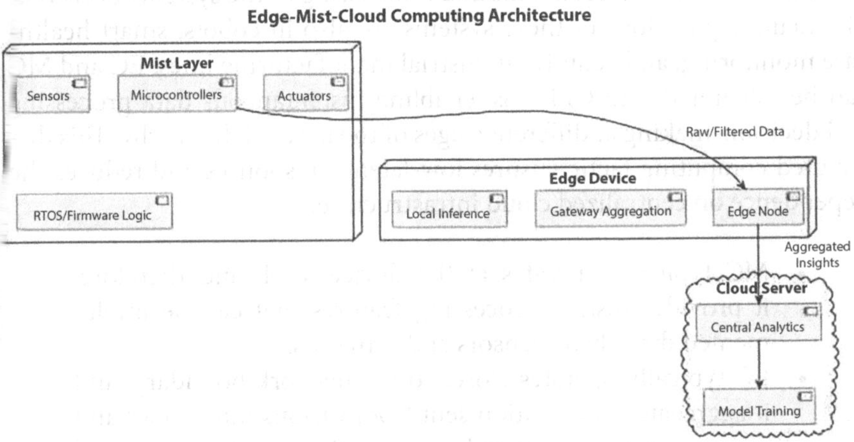

Figure 10.2 Integration architecture of EC, MC, and CC.

filtering, compression, and decision-making. This reduces the overall energy required to process data in the context of industrial systems and therefore leads to sustainability [17, 19].

Furthermore, Industry 5.0 focuses on human–machine collaboration, meaning that machines would be intelligent enough to recognize, adapt to, and collaborate with humans. This is possible in MC because MC deals mostly with real-time IoT devices. In the industry context, smartbands or devices can be used to alert users when a situation is unsafe, and edge nodes and devices can be used to send these data to management in general. Figure 10.2 portrays the integration architecture of MC, EC, and CC.

10.5 Cyberphysical Human-Centered Systems

This section describes cyberphysical human-centered systems (CPHCSs) in a detailed manner and how they can be integrated with MC and EC. A CPHS is an evolution over traditional CPSs used in Industry 5.0 applications. The main evolutionary aspect here is that it incorporates human factors and seeks to ensure that human cooperation is maintained in the system.

A CPHS represents an evolution of traditional CPSs by incorporating human elements into the core of the system. In Industry 5.0, a CPHS aims to coordinate teamwork among humans and machines. This collaboration increases productivity, safety, and adaptability in industrial environments. In CPHCSs, humans are the most important and active part of the control loop, and this allows for better flexibility and interaction, as well as providing ethical interpretations and monitoring of the system's decisions. The main applications of these systems are also in cobots, smart healthcare monitoring, and adaptive industrial manufacturing lines. EC and MC can be integrated into CPHCSs, enabling instantaneous data processing and decision-making at different stages of the network hierarchy. This distributed computing tactic ensures low-latency responses and reduces the dependence on centralized cloud infrastructure.

- MC typically operates at the device level, and therefore, it provides instant processing features that can be implemented directly on sensors and actuators.
- EC typically operates closer to the network boundary, and it aggregates information sent from various mist nodes and can perform more complex computing on the aggregated data before contacting or updating cloud services.

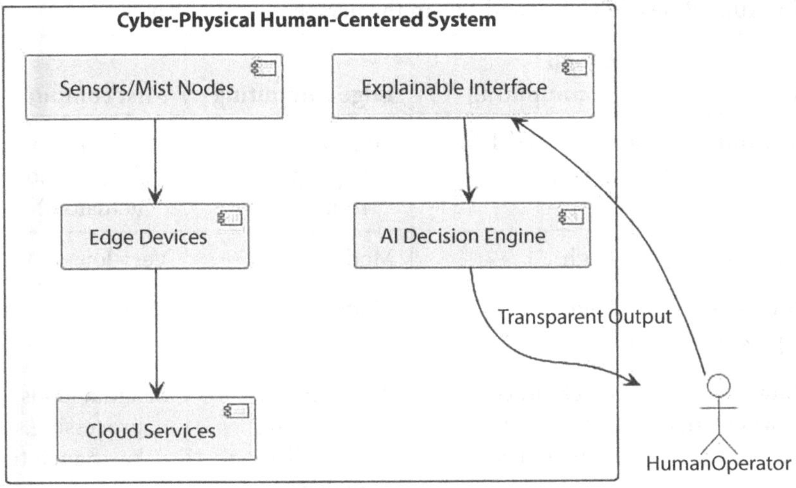

Figure 10.3 The architecture of CPHCS core components.

Figure 10.3 depicts the CPHCS core components. The CPHCS core components enhance the overall responsiveness and resilience of the CPHCS, especially in situations that require immediate human–machine interaction. Table 10.3 illustrates the role of EC, CC, and MC in the perspective of CPSs. Figure 10.4 depicts the integration of EC and MC with the CPHCS.

10.5.1 Applications of CPHCS

CPHCSs have various applications when integrated with EC and MC. This section deals with some of the major applications of CPHCSs in the context of various domains.

(i) **Healthcare**: In the healthcare context, data are private. Real-time patient monitoring, rapid diagnostics, and personalized treatment are required. EC and MC can be used in combination with CPHCSs. One such example is wearable devices, which are equipped with MC capabilities and used as a mist layer. The mist layer can now be used to process vital signs and patterns locally, ensuring that anomalies can be treated immediately. Edge nodes can be used to aggregate data from mist devices. Aggregated data can also be utilized for further examination and decision-making. Aggregated data can enhance the quality of healthcare and also increase operational efficiency [19].
(ii) **Manufacturing**: Manufacturing industries benefit from the integration of EC and MC. This integration improves the automation process.

Table 10.3 Role of CC, EC, and MC in the perspective of CPHCS.

Aspect	Cloud computing	Edge computing	Mist computing
Position	Centralized data centers	Edge of network (e.g., gateways, routers)	On-device (e.g., sensors, actuators)
Latency	High	Moderate	Very low
Processing power	High	Moderate to high	Low
Data processing	Batch processing, long-term analytics	Real-time processing, immediate analytics	Instantaneous processing, basic analytics
Connectivity dependence	Requires constant connectivity	Operates with intermittent connectivity	Operates independently of connectivity
Use cases	Historical data analysis, large-scale simulations	Real-time monitoring, predictive maintenance	Immediate response systems, safety mechanisms
Human interaction	Indirect (e.g., through dashboards)	Direct (e.g., operator interfaces)	Immediate (e.g., wearable devices, safety alerts)
Example applications	Supply chain optimization, enterprise resource planning	Smart manufacturing, autonomous vehicles	Wearable health monitors, robotic grippers

The integration framework is used to ensure predictive maintenance as well as improve safety protocols in various industries. The use of EC and MC enables manufacturing industries to have real-time monitoring of machines, which aims at immediate actions and adjustments in real time during production on production lines. MC is very important in the context of manufacturing industries because it allows the integration of many sensors and actuators, which provide a fast response to any

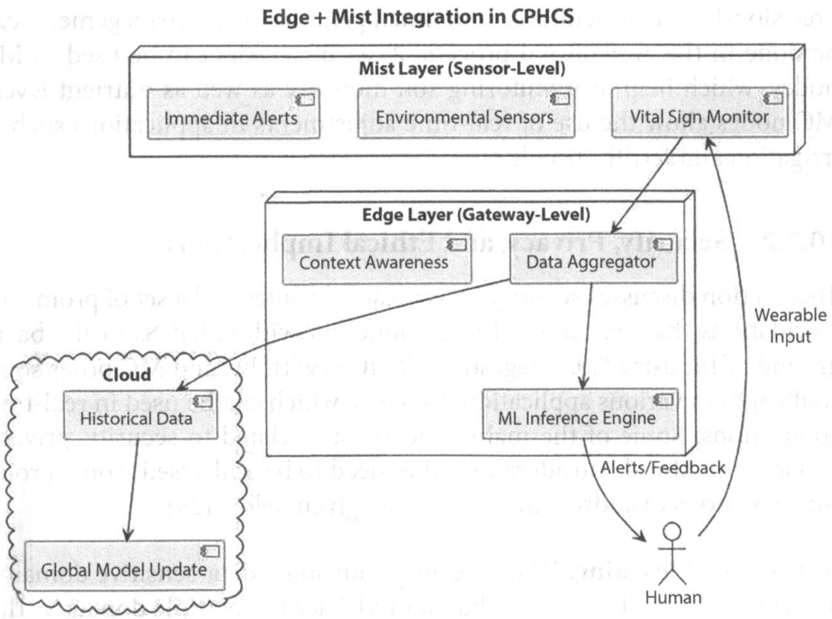

Figure 10.4 Integration of edge and mist with CPHCS.

environmental changes. MC helps in reducing the overall downtime and ultimately helps in increasing the quality of the product [20].

(iii) **Finance**: In the finance domain, EC and MC can be used for high-frequency real-time trading, fraud detection, and personalized customer service. Because computation takes place close to where the data are generated, financial institutions and organizations can reduce overall latency, increase security, and provide real-time analytics, which lead to more informed decision-making and also help improve customer experiences [21].

(iv) **Transportation**: Transportation systems can be designed to utilize the CPHCS framework for applications in real-time traffic management, autonomous vehicle operations, and even logistics. As mentioned in the above section, EC permits data to be managed nearer to the data source. In the context of traffic systems, traffic cameras, and vehicle sensors help manage traffic flow efficiently without any latency because no data are directed to the cloud for handling, thereby reducing overall latency. MC inside vehicles allows for real-time, instantaneous responses to the user based on driving conditions and can be used to improve the driving experience as well as reduce overall congestion [22].

(v) **Agriculture**: In the context of agriculture-based applications, EC and MC can be implemented. This deployment ensures that crop monitoring, precision-based agriculture, and intelligent resource management can be done in the agricultural process. A set of sensors can be used as MC nodes, which help in monitoring soil moisture as well as nutrient levels. MC nodes allow the use of real-time adjustments in applications such as irrigation and fertilization [23].

10.5.2 Security, Privacy, and Ethical Implications

This section discusses security and privacy disputes and a set of promising elucidations that can be used in conjunction with CPHCSs in the background of Industry 5.0. Integrating CPHCSs with EC and MC poses some challenges in various application domains, which can be used in real-time applications. Some of the main concerns are related to security, privacy, and a set of ethical considerations that need to be addressed. Some promising solutions to address these issues are given below [24].

(i) Federated learning: When dealing with some data-sensitive domains, it becomes difficult to train a shared model across multiple domains. This may be due to the data being very sensitive and the lack of trust in other devices or parties who want to share their private data to compute and train the ML model. In such cases, the conventional method, in which data are aggregated from clients and stored on a central server or cloud where the ML model is trained, cannot be used. FL can be used in cases where data cannot be shared between different devices; instead, model updates are aggregated in a decentralized style, which ensures that data secrecy and safety are maintained and also helps in reducing the overall communication overhead. Table 10.4 provides key differences between FL and traditional ML in a detailed manner.

10.5.3 XAI in CPHCS

This section provides a brief introduction to XAI and explains how it can be used to increase trust and transparency in the context of CPHCSs. As AI can be integrated into CPHCSs through the use of FL and other approaches, it is important to understand and make the transparency and explainability of AI decisions available to the user. XAI refers to the mixture of various ML technologies and elements used to make AI-based decisions more understandable to humans. Whereas traditional AI-based systems are typically black boxes in nature, meaning there is no way to know how the model reached certain decisions, XAI can be used to guarantee that the AI model can deliver insights into how and why a particular decision

Table 10.4 Difference between FL and traditional ML.

Aspect	FL	Centralized ML
Data location	Data stay on local devices; only model updates are shared [4].	Data are composed and stored on a central server [5].
Privacy	Improved privacy because raw data never leave local devices [4].	Data are vulnerable to breaches because all data are centralized [5].
Computation	Computation is decentralized across user devices, which may vary in capability.	Computation is centralized, often using high-powered servers.
Communication overhead	Frequent communication is needed to share model updates, leading to potential bandwidth issues.	Data are transmitted to the central server only once, resulting in less frequent communication.
Data diversity	Handles diverse, nonuniform data across devices (non-IID data).	Data are usually preprocessed and uniform when stored centrally.
Scalability	Scales well with more devices but adds complexity to managing them.	Scales by adding more computing resources to the central server.
Security risks	Vulnerable to model poisoning and inference attacks during updates.	Centralized security risks, such as data breaches.
Energy efficiency	Can drain battery and resources on edge devices.	Energy costs are concentrated on central servers.

was made, which is crucial in Industry 5.0, where human oversight, trust, and integration are required.

The following explains how XAI solutions can be integrated with Industry 5.0 applications in a detailed manner.

(a) **Smart manufacturing:** Operators working with cobots and smart systems can use XAI to create a set of dashboards that help them understand the logic behind certain outputs generated by AI, such as anomaly detection applications, predictive maintenance alerts, and quality control failures.

(b) **Healthcare:** In the healthcare context, XAI can be used to provide proper reasoning and rationale for a set of predictions made by an ML model or AI model, allowing doctors and clinicians to directly verify or challenge the result before taking any action.

(c) **Smart grids:** In power distribution systems, utility operators can use XAI to understand load-balancing decisions and predict anomalies in real time.

Table 10.5 Commonly used XAI techniques in various applications.

Technique	Description	Suitable for CPHCS layer	Example application
LIME	Explains predictions of any model locally	Edge/cloud	Real-time maintenance alerts
SHAP	Uses Shapley values for feature importance	Edge/cloud	Healthcare diagnostic decisions
Decision trees	Simple, interpretable decision paths	Mist/edge	Wearable health monitor logic
Generalized additive models	Additive models with visual interpretability	Edge/cloud	Financial risk analysis
Saliency maps	Visual input attribution for deep learning models	Cloud	Medical imaging/ vision systems
Counterfactual explanations	Provides minimal input changes for different outputs	Edge/cloud	Smart grid load decisions
Rule-based systems	Transparent if–then rules	Mist/edge	Factory safety shut-off controls

The above deals with some of the main applications of XAI in the framework of CPHCS and Industry 5.0. Table 10.5 briefly describes the commonly used XAI methods in various applications and what they do in the context of XAI.

Despite the many benefits such as transparency and interpretability, there are some challenges that need to be overcome in the context of XAI implementation, one of which is usually the accuracy-and-interpretability trade-off, i.e., the more accurate the model, the less interpretable it is, which requires the user to balance accuracy and transparency during ML model design. Another challenge in the context of scalability is that, due to resource constraints in EC and MC, the interpretations need to be scaled to work in a distributed infrastructure.

10 6 Conclusion

This chapter focuses on integrating advanced computing innovations such as EC, MC, and CC with CPHCSs. This chapter discusses how this integration can help the transformation from the digital focus of Industry 4.0 to the human-centric Industry 5.0 by concentrating on human–AI collaboration, ethical design approaches for AI, and creating and maintaining sustainable systems. In general, the integration of EC and MC provides decentralized and real-time computing capabilities, and its integration with CPHCSs helps human–AI collaboration. Together, these technologies not only increase efficiency and responsiveness but also enhance trust, integrity, and contextual awareness, especially in sensitive domains such as healthcare. Future improvements and studies are currently focused on the strategic integration of these components with industries, which will help build robust intelligent systems in real time, which are also aligned with human emotions, thus making them truly Industry 5.0 in nature.

AI disclosure statement: The author confirms that no artificial intelligence (AI)-generated content or AI-assisted writing tools were used in the preparation of this chapter (including summaries and the abstract). All text, analysis, and conclusions are the result of the author's original work.

References

1. Kagermann, H. and Wahlster, W., Ten years of Industrie 4.0. *Sci.*, 4, 26, 2022.
2. Lee, J., Bagheri, B., Kao, H.A., A cyber-physical systems architecture for industry 4.0-based manufacturing systems. *Manuf. Lett.*, 3, 18, 2015.
3. Ivanov, D., The Industry 5.0 framework: viability-based integration of the resilience, sustainability, and human-centricity perspectives. *Int. J. Prod. Res.*, 61, 1683, 2023.
4. Sung, T.K., Industry 4.0: a Korean perspective. *Technol. Forecast. Soc. Change*, 132, 40, 2018.
5. Shi, W., Cao, J., Zhang, Q., Li, Y., Xu, L., Edge computing: Vision and challenges. *IEEE Internet Things J.*, 3, 637, 2016.
6. Satyanarayanan, M., The emergence of edge computing. *Comput.*, 50, 30, 2017.
7. Satyanarayanan, M., Bahl, P., Caceres, R., Davies, N., The case for VM-based cloudlets in mobile computing. *IEEE Pervasive Comput.*, 8, 14, 2009.
8. Bonomi, F., Milito, R., Zhu, J., Addepalli, S., Fog computing and its role in the Internet of Things, in: *Proc. of the MCC Workshop on Mob. Cloud Comput.*, Finland, 1, 13, 2012.
9. Chiang, M. and Zhang, T., Fog and IoT: An overview of research opportunities. *IEEE Internet Things J.*, 3, 854, 2016.
10. Tomovic, S., Yoshigoe, K., Maljevic, I., Radusinovic, I., Software-defined fog network architecture for IoT. *Wirel. Pers. Commun.*, 92, 181, 2017.
11. Computing, E.M.E. and Initiative, I., Mobile-edge computing: an introductory technical white paper, in: *Proc. ETSI: Sophia Antipolis*, France, 1, 36, 2014.
12. Bilal, K., Khalid, O., Erbad, A., Khan, S.U., Potentials, trends, and prospects in edge technologies: Fog, cloudlet, mobile edge, and micro data centers. *J. Comput. Netw.*, 130, 94, 2018.
13. Gubbi, J., Buyya, R., Marusic, S., Palaniswami, M., Internet of Things (IoT): A vision, architectural elements, and future directions. *Future Gener. Comput. Syst.*, 29, 1645, 2013.
14. Jonas, E., Schleier-Smith, J., Sreekanti, V., Tsai, C.C., Khandelwal, A., Pu, Q., Patterson, D.A., Cloud programming simplified: A Berkeley view on serverless computing. *arXiv preprint arXiv:1902.03383*, 2019.
15. Lane, N.D. and Georgiev, P., Can deep learning revolutionize mobile sensing. *In Proc. of the 16th Int. Workshop on Mob. Comput. Sys and App.*, United States, 1, 117, 2015.
16. Macías, J.E.Z. and Arlucea, A.A. (Ed.), *Integrative Cloud to Mist Computing: Architectures, Applications, and Innovations*, Data Engineering. Rolando Herrero, 2016.
17. Nahavandi, S., Industry 5.0 - A human-centric solution. *Sustainability*, 11, 4371, 2019.

18. Radanliev, P., De Roure, D., Page, K., Nurse, J.R., Mantilla Montalvo, R., Santos, O., Burnap, P., Cyber risk at the edge: current and future trends on cyber risk analytics and artificial intelligence in the industrial Internet of things and industry 4.0 supply chains. *Cybersecurity*, 3, 21, 2020.
19. Gomathi, L., Mishra, A.K., Tyagi, A.K., Industry 5.0 for healthcare 5.0: Opportunities, challenges, and future research possibilities. *In Int. Con. on Trends in Electronics and Info.*, India, 7, 204, 2023.
20. Sharma, M., Tomar, A., Hazra, A., Edge computing for industry 5.0: Fundamental, applications and research challenges. *IEEE Internet Things J.*, 11, 71, 2024.
21. Lakshminarayanan, V., Ravikumar, A., Sriraman, H., Alla, S., Chattu, V.K., Health care equity through intelligent edge computing and augmented reality/virtual reality: a systematic review. *J. Multidiscip. Healthc.*, 16, 2839, 2023.
22. Zhou, X., Ke, R., Yang, H., Liu, C., When intelligent transportation systems sensing meets edge computing: Vision and challenges. *Appl. Sci.*, 11, 9680, 2021.
23 Ayaz, M., Ammad-Uddin, M., Sharif, Z., Mansour, A., Aggoune, E.H.M., Internet of Things (IoT) - based smart agriculture: Toward making the fields talk. *IEEE Access*, 7, 129551, 2019.
24. Kumar, E.S., Ramamoorthy, R., Kesavan, S., Shobha, T., Patil, S., Vighneshwari, B., Comparative study and analysis of cloud container technology, in: *Proc. Int. Con. on Comput. for Sust Global Dev.*, India, 11, 1681, 2024.

11

An Intelligent Cyberphysical System Threat Detection for Secured Communication in 6G Autonomous Vehicle Networks

N. Sathish[1]*, A. Prasanth[2], V. Yokesh[3] and Afizan Bin Azman[4]

[1]*Department of Electronics and Communication Engineering, Sri Venkateswara College of Engineering, Sriperumbudur, Tamil Nadu, India*
[2]*Department of Computer Science and Engineering, Vel Tech Rangarajan Dr. Sagunthala R&D Institute of Science and Technology, Chennai, Tamil Nadu, India*
[3]*Department of ECE, Sathyabama Institute of Science and Technology, Chennai, Tamil Nadu, India*
[4]*Faculty of Innovation & Technology, Taylor's University, Subang Jaya, Selangor, Malaysia*

Abstract
Autonomous vehicle ecosystems are being completely transformed by the rapid development of 6G networks, which make seamless interconnection and ultrareliable, low-latency communication possible. Because autonomous vehicle networks are built on cyberphysical systems (CPSs), secure communication is essential to thwarting cyber threats such as advanced persistent threats, spoofing, denial-of-service assaults, and data breaches. The dynamic and intricate threat landscape associated with 6G-enabled autonomous vehicles is difficult for traditional security methods to handle. An intelligent threat detection methodology for CPSs in 6G autonomous vehicle networks is presented in this book chapter. It investigates how to combine deep learning, machine learning, and artificial intelligence models for secure data transfer, intrusion prevention, and real-time anomaly detection. It also covers how federated learning, blockchain, and quantum-safe cryptography might improve security, privacy, and trust. This chapter offers insights into the creation of robust and intelligent security solutions for upcoming autonomous

**Corresponding author*: n.satyish@gmail.com; ORCID: 0009-0000-3507-6337

Anitha Velu, Prasanth Aruchamy, Raghu Ramamoorthy, Rajesh Kumar Dhanaraj and Manjula Arunraj (eds.) Cyber Physical Human Systems: Emerging Trends and Applications, (249–270) © 2026 Scrivener Publishing LLC

transportation systems by tackling the changing cybersecurity issues in next-generation vehicular networks.

Keywords: Autonomous vehicle networks, cybersecurity, 6G communication, block chain, artificial intelligence, machine learning, deep learning

11.1 Cyberphysical Systems in 6G Autonomous Vehicle Networks

The development of communication technologies has never ceased to serve as the key point in determining the abilities of autonomous systems. CPSs are one of the most transformational creations in the recent past that connect computing algorithms with physical systems to allow real-time communication between the real and virtual worlds [1]. In self-driving car applications, a CPS is the spine that allows such cars to be aware of the world around them and behave in real time by making real-time decisions and automating navigation without human involvement. Due to the anticipated ultralow latency, high-bandwidth, and high-density interconnection of 6G networks, the performance and reliability of CPSs are to be further enhanced, enabling the next generation of AV networks. The chapter discusses the integration of CPSs in the area of AVs, the revolutionary nature of 6G on CPSs, and how it would eventually culminate to produce safer, more efficient, and smarter transportation systems in the near future.

11.1.1 Introduction to Cyberphysical Systems

The formation of the links between the digital space and the physical one generates a CPS as the integration of the computational element with the surrounding physical environment into a holistic construct. The central components of a CPS are the interplays between embedded systems, sensors, actuators, and the computational algorithms on that the decision-making process depends [2]. Not only must such systems use real-time sensor information to interact with their environment, but they also must have a very responsive and distributed computational environment to process the information and make sure the physical parts are operating as planned. The range of applications of CPSs includes robotics used in the industry as well as smart grids, yet few areas can enjoy the potential of this technology more than AVs.

A CPS of autonomous automobiles is a combination of hardware and software in which the sensors incorporated in the automobile repeatedly scan the vehicle environment and the onboard processing system examines and processes the information. These systems must respond to environmental stimuli in real time, be it identifying pedestrians, assessing road conditions, or computing the best paths. In AVs, labeling a CPS by key features is the following: operating in real time, feedback control, and continuous controlling and regulation of car functions. The features are crucial to the safety, efficiency, and reliability of autonomous cars, which are required to drive in complex and unpredictable environments.

11.1 2 Role of CPS in Autonomous Vehicles

AVs use a CPS as a fundamental element in their capability. A CPS can be viewed at each of the operations of an AV, which includes perception and decision-making as well as control and navigation [3]. The key to this system is to obtain real-time data using a variety of sensors such as cameras, light detection and ranging (LiDAR), radar, global positioning system (GPS), and ultrasonic sensors. The data are transferred to the computational system of the vehicle, which processes them with the help of advanced algorithms, such as AI and ML. The knowledge generated with the help of these data helps the vehicle to perceive the surrounding space correctly, comprehend the context, and make accurate decisions about path finding and security.

Real-time perception is one of the key roles of a CPS in self-driving cars. The sensors on the vehicle receive signals of the environment, which includes other road users, pedestrians, road signs, road conditions, and obstructions [4]. This information is then used within the computational systems where the vehicle is allowed to construct an up-to-date map of the surroundings. With this knowledge, the car decides on acceleration, braking, and steering to prevent collisions or to whatever effect (to make the best out of the path chosen).

The CPS in self-driving cars also offers strong feedback control. The implication of this characteristic to the vehicle is that it can always adapt its behavior to the changing environmental conditions. As an example, when a pedestrian runs across the road, the vehicle will make a fast decision, apply brakes, and can alter the steering within a very short time. The exchange between the physical and cyber constituents makes the vehicle have a balance between safety and efficiency. Also, the CPS is compatible with vehicle-to-vehicle (V2V) and vehicle-to-infrastructure (V2I) communications, which enable cooperation with other cars and infrastructure, such as traffic lights, road signs, and smart cities.

11.1.3 Impact of 6G on CPS Integration and Performance

The advent of 6G networks promises to transform how CPSs will perform, especially when considering AVs. 6G is the successor of wireless communication that will offer ultralow latency, huge device connection capacity, and speeds of data transfer much beyond what is expected of the current 5G telecommunication networks. In the case of autonomous cars, the introduction of 6G-based CPSs is going to feature unmatched benefits in the areas of data communication, system reliability, and responsiveness [5]. Among the greatest contributions of 6G to CPSs to AVs, it is possible to identify real-time communications improvements. Self-driving cars will depend on the constant flow of information provided by their vehicle sensors and other networks (autonomous cars and infrastructure). 6G networks will be able to transmit this information faster and more reliably, and thus, within the shortest time possible. As an illustration, a 6G-empowered car will be able to update the road conditions, receive notifications regarding traffic and hazards, as well as improve awareness of the situation and react to it more quickly as it will communicate with other vehicles or city infrastructure in real time.

In addition, the scalability and robustness of the CPS infrastructure will also be increased as 6G will allow an enormous number of devices to be connected to it. In a heavily populated urban setting and when numerous cars, pedestrians, and intelligent devices have just gotten connected, 6G will support smooth communications between a plethora of devices. It will enable increased syncing of the vehicles, infrastructure, and people, with fewer people involved in the whole traffic flow, being safer and energy-efficient. Also, the increased data rates and improved signal quality provided by 6G will allow self-driving cars to work with more data more efficiently, enhancing their sense and judgment.

Moreover, 6G will most likely be capable of intelligent networking support and optimization with the use of AI. Advanced AI and ML algorithms together with 6G will enable vehicles to predict and optimize the traffic patterns, better route, and even learn to drive by the behavior of the other cars. Such a 6G-enabled decision-making AI tool will produce more effective and safer networks of autonomous cars.

Greater reliability of the system provided by 6G will also be very important in the safety of self-driving cars. Because the vehicles will have access to a more stable, secure, and fault-tolerant set of communication networks, the chances of power shutdown because of poor connectivity or even clogging of the network will be greatly reduced. The high performance of a CPS along with the innovation of 6G will help in providing high resiliency of the AV, especially in the most vulnerable places such as urban intersections, highways, and in poor weather conditions.

The possibility of CPSs' fusion of 6G to the networks of AVs has enormous potential in terms of the development of vehicle autonomy. Self-driving cars will become much more functional and safer due to the high-speed, ultrareliable communication functionality of 6G and the support of extremely high data and device quantities. Such integration will lead to a future where AVs can be smarter, safer, and more efficient in moving through a networked and data-centered world.

11.2 Security Challenges in 6G-Enabled CPS for Autonomous Vehicles

The convergence of a CPS in autonomous cars has transformed the world of transportation in that cars can be driven virtually without the driver as they still interact with the physical world and the cyber environment [6]. But as the 6G technology has emerged, which is capable of providing high-speed connectivity, ultralow latency, and improved interconnectivity of the devices, there emerges a new set of security challenges that are quite complicated. These issues have to be dealt with so as to guarantee that the AVs can operate securely and safely by using CPSs extensively. This chapter elaborates on the security threats, 6G-enabled CPSs in AVs, the evolving threat scenario, the inadequacies of the traditional security measures, and real-time distributed threat dynamics at play in such interconnected systems.

11.2.1 Threat Landscape: APTs, Spoofing, DoS, and Data Breaches

Emerging complexity and interconnectivity of self-driving vehicles controlled by CPSs in a 6G network introduce new possibilities of attack vectors, which malicious players might use. Without a comprehension of the threat situation, it is impossible to develop effective countermeasures.

(i) Advanced Persistent Threats

APTs are among the most hazardous types of breaches that can be used against AVs. These are long-term, advanced, and covert attacks in which criminals in the computing world hack into computer systems and stay there in the long term. As far as CPS-enabled self-driving cars are concerned, an APT might entail unauthorized access to the car onboard electronic features, such as navigation, communication, and safety systems [7]. The attackers might take advantage of the APTs to gain control over the vehicle to act in a dangerous way that puts the lives of passengers and others at risk of accidents. As an example, the hacker can enter the sensor network of the vehicle and alter the data flows, feeding the vehicle with wrong data forcing it to make erroneous judgments.

(ii) Spoofing

Spoofing entails deceiving a system that it is communicating with a genuine system yet it is communicating with a villain. This is especially worrying with regard to an autonomous car that heavily depends on data accuracy provided by different sensors (such as GPS, radar, or LiDAR). The fake signals might be sent to spoof navigating systems of the vehicle. There is a risk of a malicious actor taking on the appearance of the GPS satellites or producing fake road signs, causing the vehicle to go down the wrong paths or not recognize obstructions. Spoofing attacks will become more advanced as connected cars attend more to real-time data distribution or sensor data [8].

(iii) Denial of Service Attacks

DoS and distributed DoS are the types of attacks that aim to cripple the capacity of the system to become unusable. Considering the scenario of AVs, a DoS may be conducted against the vehicle systems of computation or communication channels; it may affect prompt processing of sensor data, V2V networks, or V2I networks data [9]. The consequence of DoS attacks in a 6G-based CPS where ultrafast communication and real-time interaction is essential may be catastrophic. As an example, hackers can overload the system with too much traffic, hence causing delay or even losing some data, thus making the wrong decision, and that can result in accidents.

(iv) Data Breaches

Because AVs can produce enormous data, including sensitive personal data, location, behavior patterns, and travel schedules, they will be excellent data breach targets. Attackers may use vulnerabilities in the 6G network to eavesdrop and steal any data being transmitted or hack the vehicle memory stores. Besides being a privacy risk, these violations may also cause legal and financial repercussions to vehicle manufacturers, service providers, and users [10]. Besides, the stolen data may be deployed to track people, conduct social engineering attacks, or manipulate a vehicle remotely.

11.2.2 Limitations of Traditional Security Mechanisms

The current security mechanisms that have been used to secure traditional networks include firewalls, encryption mechanisms, and authentication procedures. These conventional methods, however, have a number of limitations when it comes to the context of 6G enablement of a CPS that is being applied to the AVs.

(i) Scalability Issues
The inability to scale appropriately within the hyperconnected 6G environment is one of the biggest weaknesses of traditional security mechanisms. There are millions of devices, sensors, and vehicles that will always be in constant communication in a 6G-enabled self-driving vehicle network. These systems are also characterized by massive scale requirements and require real-time responses, thus traditional security solutions such as the centralized authentication systems or centralized access control systems can prove difficult in managing these requirements successfully. To achieve security on this scale, much more adaptive and decentralized methods can react to threats on the scale of these vast and distributed networks [11].

(ii) Latency and Real-Time Constraints
The classic security types usually induce latency, and this aspect can be an inconvenience in AVs with real-time decision-making. To give an example, encryption procedures or complicated multifactor authentication procedures can interrupt transmission of important sensor data, e.g., road conditions, traffic lights, or obstacle detection. At even a millisecond of delay in processing, accidents can take place in the AV. Hence, the classic security mechanisms where the focus is to implement static security measures might not satisfy the low latency requirements of the 6G networks used within CPS setups [12].

(iii) Lack of Context-Aware Security
The conception of the security systems is usually performed to respond to some previously established threats and patterns, which may decrease the chances to adjust to the new and changing attack situations. Nevertheless, threats of the 6G-enabled CPS of AVs are rather contextualized and changing continuously. As an example, the surrounding of the vehicle is continually evolving, and the same must happen with the security standards. To respond to the emerging threats in a proactive, flexible way, context-aware security mechanisms are required that can utilize real-time data,

vehicle status, and environmental information to dynamically adjust security protocols to take advantage of the flexibility. This kind of flexibility cannot be found in the traditional security schemes.

(iv) Vulnerabilities in Supply Chains and Software
Conventional security systems are keen on addressing short-term cybersecurity of gadgets, networks, and systems, and adequate attention is not paid to the chain of supply and software susceptibility in AV systems. Malicious actors may execute security gaps to be produced in the process of manufacturing, software updates, or even the time of integrating a third-party software into the AV. As an example, a bug in any third-party software used during communication or navigation may affect the whole car. End-to-end security audits and closer monitoring integration of threats may be required to proceed with traditional security procedures, which might not be ready to deal with such an attack.

11.3 Real-Time and Distributed Threat Dynamics

The real-time dynamics of threats and distributed attacks are further complicated when 6G technology and CPSs are used to control AVs. Self-driving will occur by being linked to a network of interconnected devices, the road infrastructure, and the services that exist in the cloud, and all these are required to run in real time [13].

11.3.1 Threat Detection and Response in Real Time

Threats can be raised on demand, in real time in a 6G-enabled CPS, and thus detection and response actions must be real-time. A hacker may exploit the V2V, V2I, and vehicle-to-cloud communication connections of the vehicle, and by the time that traditional detection devices detect the anomaly, it may be too late to avoid a serious accident or breach the cybersecurity of the vehicle. In real-time threat detection, there is a need to have adaptive and AI-based systems that have the ability to analyze the incoming data streams to detect any malicious activity much faster than traditional approaches permit. Considering an example whereby the vehicle is saturated with fake GPS signals, the real-time detection solution should determine the deviation and compare it against other sensor records (such as radar or LiDAR) and generate an alarm or consideration. The untraditional nature of a CPS is such that any conventional threat response mechanism cannot track it properly: in a CPS, early is the difference between a safe event and an accident, so the handling mechanisms must be quicker than ever before.

11.3.2 Additional Dynamics of Distributed Attacks in the 6G Ecosystem

Distributed 6G networks imply that one attack may spread rapidly along several vehicles or nodes, increasing its effects. The security of a vehicle or device is quite solitary in old systems. Yet, in a 6G-enabled CPS, there will be V2V or V2I communication that will be able to spread attacks. This forms a distributed attack plane in the sense that flaws on one automobile or infrastructure node can spread to the whole network [14]. An example is a DoS attack on the 6G Internet infrastructure that may create havoc in urban centers, including not only AVs but also traffic lights and signage, smart road and road safety infrastructure, and emergency codes. There is a possibility of attackers using botnets or hacked vehicles to launch such attacks, and this makes the threat hard to isolate and address. Distributed cybersecurity frameworks will have to support the integrity of distributed systems, and a new generation of these frameworks will need to be able to identify and react to threats throughout a large, interdependent ecosystem of AVs and infrastructure.

11.4 Intelligent Threat Detection Framework

The dynamic nature of cyberattacks requires an intelligent and adaptive threat detection framework to secure the 6G-enabled AV networks. The classic security models are inadequate in dealing with the real-time, high-speed, and dispersed nature of CPSs. As such, the suggested framework will make use of a mixed approach based on AI, ML, and highly advanced communication technologies, allowing for proactive detection of threats, prompt mitigation, and real-time system adaptation. The subsections below outline the architectural design, main building blocks, and the life cycle of threat detection that is designed with intelligent autonomous transportation environments in mind [15].

11.4.1 Architecture of the Threat Detection System

The suggested layout of the automated threat detection solution for 6G-powered AV networks is a multitier one, co-opting the use of a CPS and smart AI-based safeguarding features. Fundamentally, the architecture involves the integration of a decentralized intelligence framework anchored on edge computing nodes that give edge data processing capabilities that occur in real time and on-demand data analysis and monitoring of threats that directly happen in the vehicle or on the roadside infrastructure. The layered architecture contains the perception layer that gives raw data on the environment and within the system, a decision layer that incorporates ML and AI algorithms to determine anomalies and categorize threats, and an action layer to conduct a system response such as isolation, alert, or adaptation. The layers are interconnected by a cryptographically secured communication backbone based on the use of URLLC, which provides vehicles and infrastructure with highly synchronous communications. The architecture also comprises blockchain-based logging using an immutable ledger of threat events and updates, which enables auditability, traceability, and collaborative threat intelligence sharing throughout the vehicular network.

11.4.2 Key Components: Sensors, Actuators, and Communication Interfaces

CPS components are abundant, which enables the intelligent threat detection system, and each of the CPS components contributes immensely to sensing, interpreting, and responding to the threats. The data acquired from sensors, such as LiDAR, radar, GPS, ultrasonic detectors, and environmental cameras, gather high-fidelity, real-time information on the environment, vehicle dynamics, and interaction with the infrastructure. The sensors serve data to onboard processors and ML applications that scan the environment to identify anomalies that are a sign of spoofing, APTs, or DoS situations. The physical response level is made of actuators that perform the necessary commands (braking, steering corrections, system shutdown, etc.) following the observed threat. They are directly dependent on the results of AI-based threat analysis modules. In the meantime, interfaces of communications with vehicles (V2V, V2I, and V2X) allow sharing situational awareness and threat intelligence, creating the distributed security mesh. These interfaces are strong and can also be strengthened by the 6G high bandwidth and low latency that will be able to send security alerts and countermeasures simultaneously across the network.

11.4.3 Threat Detection Life Cycle

The threat detection life cycle in this model takes a dynamic, intelligent, and adaptive route. The threat detection life cycle is illustrated in Figure 11.1. The first step is data acquisition, which involves the collection and preprocessing of the data received at various heterogeneous sensors and also in the system logs to eliminate noise and maintain consistent data. These data are analyzed using the supervised and unsupervised learning models in the detection phase, and they find the patterns or deviations that can indicate the presence of a threat such as sensor spoofing, malevolent command injections, or DoS [16]. The system subsequently enters the evaluation stage, in which the situational awareness (e.g., vehicle speed, environmental conditions, and network status) is used to confirm the threat and to ascertain its seriousness. Subsequently, during response, the system implements the most suitable mitigation action, which includes possibly the warning to human operators, up to the involvement of automated defense capabilities, e.g., rerouting, isolation of the infected nodes, or activation of emergency controls. Lastly, during the learning and adaptation step, the system can record the event to an FL model or blockchain, which lets in the continuous updating of the threat model and also allows multiple distributed vehicles and edge nodes to learn collaboratively. Such a back-and-forth cycle guards the detection framework to be proactive, resilient, and progressively more precise with time in a continuously challenging 6G threat environment.

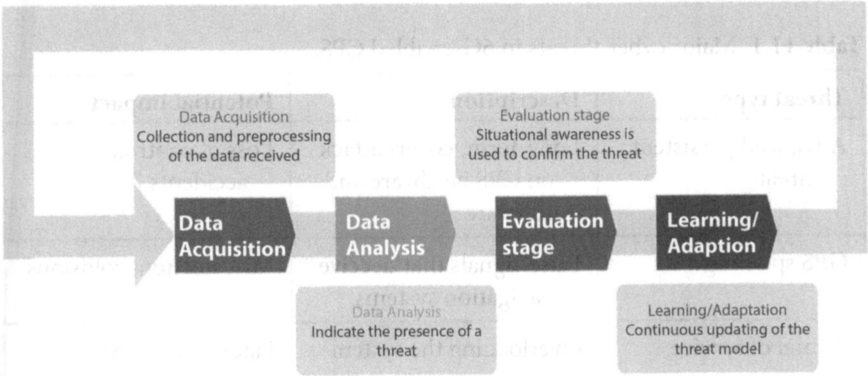

Figure 11.1 Threat detection life cycle.

11.5 Artificial Intelligence and Machine Learning for Threat Detection

The growing sophistication of cyber threats in CPSs, based on 6G technologies, requires the incorporation of innovative computational intelligence methods. AI and ML are critical when it comes to recognizing, categorizing, and reacting to malicious activity in real time, particularly when isolating self-driving vehicles [17]. With their capacity to be data-driven and enabled to learn over time using the history of attacks to refine or adjust to the changing landscape of threats, the AI-enabled framework offers an adaptive, scalable, and proactive line of defense [18]. In the subsections below, the chapter discusses how various learning models and DL architectures with real-time intrusion prevention plans are exploited to ensure that the communication between AVs is secure. The major cyber threats in 6G-enabled CPSs are listed in Table 11.1.

11.5.1 Supervised and Unsupervised Learning Models

AI-based threat detection is based on supervised and unsupervised learning models in autonomous CPS environments. Supervised learning uses labeled data, containing known threat signatures and clean data, to conduct supervised training of such model paradigms as decision trees, support vector machines, and ensemble approaches such as random forests. The known types of attacks that they can identify accurately include

Table 11.1 Major cyber threats in 6G-enabled CPS.

Threat type	Description	Potential impact
Advanced persistent threat	Long-term, covert attack on CPS hardware and software	Loss of control, accidents
GPS spoofing	False signals that deceive navigation systems	Misdirection, collisions
Denial of service	Overloading the system prevents decisions or communications	Latency, crash risk
Data breaches	Private information taken away from car systems	Privacy loss, data leaks

DoS or spoofing through learning the differences that can be noticed through the network traffic or sensors in general. On the other hand, in cases where there is limited labeled data or it is expected to encounter unknown threats, unsupervised learning methods are utilized (k-means clustering, density-based spatial clustering of applications with noise [DBSCAN], and autoencoders). These models examine data distributions and identify the deviations or anomalies that point to the occurrence of new or modified attacks that traditional rule-based systems will not be able to capture. The ability to find a compromising threat in AV ecosystems in 6G involves the creation of models of supervision and the integration of unsupervised models, which makes it possible to ensure both accuracy in the identification of threats and the plasticity in recognizing new problems.

11.5.2 Deep Learning for Anomaly Detection

DL presents multiple layered neural network models, which allow extraction of deep level features of large, complex data streams produced by AV systems. In particular, models such as convolutional neural networks, recurrent neural networks, and long short-term memory networks prove well-suited to analyzing time-series sensor data, traffic logs, and image feeds [19]. To detect anomalies, DL models may be trained on routine operation data of the vehicle and the infrastructure at large to capture a model of expected normal behavior. Any substantial deviation from this baseline—e.g., abnormal values represented by sensors, patterns of communication, and vehicle dynamics—can be marked as possible threats. Autoencoders, specifically, are helpful when trying to reconstruct the normal input signals and determining anomalies using the reconstruction error. The dynamic environment of high throughput requires models to be continuously enhanced with new data, which makes them highly suitable to the 6G environment, where real-time adaptation and learning are essential to ensuring that autonomous operations remain secure [20].

11.5.3 Real-Time Intrusion Prevention Mechanisms

Protection from intrusion in a 6G-based autonomous network will have to be done in real time, where an AI model would not only detect the threat but would have to act in real time to start mitigation actions. Edge AI is used to deploy intrusion prevention systems (IPSs) in the form of real-time IPSs where inference engines are deployed as part of onboard systems within the vehicle or to roadside infrastructure nodes so that there is minimal latency [21]. When a possible malicious activity is detected—for example, an unauthorized access attempt, an abnormal pattern of data transmission, or a malicious signal injected into the sensor input—the IPS can respond through the placement of defensive measures, such as traffic filtering and access denial or communication flow rerouting. Others have also used reinforcement learning–based systems to deal with real-time prevention, wherein the agents learn how to best respond more and more over time based on environmental feedback and past success. The advantage to these AI-driven IPSs is in FL models where the systems can learn together and cooperate across the world in threat patterns without having to share their data and compromise their privacy. The incorporation of intelligence at the prevention stage makes the response to threats quick, context-related, and seamless, protecting the safety-critical functions in AVs in a hyperconnected 6G environment.

11.6 Emerging Technologies for Secure Communication

The development of the 6G network and the emergence of AVs that work within the smart CPS make the security measures that were traditionally used inefficient to cope with the magnitude, heterogeneity, and real-time constraints of vehicular networks. To guarantee the end-to-end security of communications, new technologies including FL, blockchain, and quantum-safe cryptography are being deployed to enhance privacy, trust, and data resilience [22]. These technologies not only solve current weaknesses but also anticipate future threats in a scalable decentralized security problem that enjoys the benefits of distributed environments in a dynamic world. The sections below give a brief description of the contribution of each of these technologies toward the realization of a secure and intelligent communication infrastructure [23].

11.6.1 Federated Learning for Privacy-Preserving Security

The paradigm of FL enables a decentralized ML model, and such a configuration improves privacy because models can be trained across a set of devices or nodes, including AVs or nodes built along the roadway, without the necessity of raw data being transferred to a centralized server [24]. Upon the premises of 6G-based vehicle networks, FL is used in privacy-friendly intrusion detection so that each vehicle can learn in its immediate environment and yet provide model updates into a global threat detection model.

This will guarantee that sensitive information such as personal travel routes, behavioral patterns, or the data from environmental sensors will not find itself on exterior systems but will be left on the device. In addition to that, FL can improve model generalization by means of amalgamation with various settings, translating to stronger and flexible security models. Integrated with differential privacy and secure aggregation strategies, FL is an efficient mechanism to avoid data leakage and protect sensitive vehicular activities against cyberattacks without affecting detection performance.

11.6.2 Blockchain for Trust and Data Integrity

Blockchain technology introduces decentralized trust and tamper-proof record keeping at the center of secure car communication. Within an autonomous transportation network that incorporates 6G connectivity, blockchain may be utilized to generate tamper-free transactional records, sensor data transfers, and threat intelligence broadcasts throughout a decentralized network of vehicles and network devices. This way, a record of a single piece of information can never be modified or fabricated, whether it is a threat alert, vehicle status, or an environmental observation, guaranteeing a source of truth when reaching a decision together. Moreover, it is possible to implement smart contracts so that automated, rule-based security actions or access controls are implemented [25]. As an example, an attribution strategy based on blockchain technology might be able to authenticate identity and capture the history of a vehicle before providing access to vital road resources or confidential data. Such a distributed ledger mechanism bypasses the use of a central authority, avoids excessive single-point-of-failure risk, and exudes confidence between heterogeneous nodes within the autonomous system.

11.6.3 Quantum-Safe Cryptography for Future-Proofing

Quantum computing is a major threat to the traditional cryptographic algorithms that are deployed to secure automated vehicular communications as it evolves. The embedding of quantum-safe cryptography, or postquantum cryptography, is increasingly being treated as a necessity in ensuring the future-proofing of 6G-enabled CPS systems and self-driving cars. These quantum cryptography algorithms are developed to counter the computing capabilities of quantum computers and provide long-term safety to sensitive information including user authentication keys, navigation and control signals, and the V2I interface protocols [26]. The most common candidates that are being considered to be implemented in the vehicular environment consist of lattice-based cryptography, hash-based signatures, and multivariate polynomial schemes. Using quantum-resistant cryptography makes it impossible to decode the encrypted data even in case they have been intercepted today; they will remain indecipherable even when quantum computers become widespread. This will ensure that the confidentiality and integrity of AV systems will be preserved over time, and quantum-safe cryptography will be an important element in a robust and secure communication infrastructure in the next generation of intelligent transport systems.

11.7 Integration of Intelligent Systems in 6G Autonomous Networks

The convergence of intelligent systems and 6G communication infrastructure is pivotal to enabling seamless, secure, and responsive AV operations. As vehicles become smarter and more interconnected, the underlying CPS must efficiently manage vast data streams, real-time decision-making, and coordination across distributed network components. The integration of AI, edge intelligence, and context-aware communication frameworks must be balanced with the need for low latency, high reliability, and scalable system interoperability. This section explores how emerging techniques such as interoperability standards, edge computing, and network slicing contribute to enhancing performance, responsiveness, and customization in 6G-based autonomous vehicular networks.

11.7.1 System Interoperability and Latency Considerations

System interoperability is essential in the implementation of reliable communication among diverse devices, platforms, and communication protocols in 6G-enabled AV ecosystems. Automobiles are usually designed with different hardware structures, operating systems, and vendor interfaces, which may cause fragmentation in information sharing and response to threats. The key to having smooth interoperation is having standard protocols of communication, real-time APIs, and shared data models through which a harmonization of decisions can easily be made in a multivendor environment. Meanwhile, in autonomous driving, milliseconds may be the difference between safe passage and an accident, so the ultralow-latency capabilities of 6G, targeting less than a millisecond of delay, can support V2V and V2I types of communications in real time. In mission-critical situations such as avoiding obstacles or road signals and emergency braking, intelligent systems need to be incorporated in a way that does not lose this responsiveness. It is therefore necessary to strike a balance between compatible systems and having low-latency processing capabilities required to facilitate effective AV operation in the smart city.

11.7.2 Edge Computing and Distributed Processing

Edge computing has been at the forefront in the integration of intelligent systems by shifting data processing and analytics near the source—i.e., inside or near the vehicle or infrastructure that is processing raw data—instead of it being sent to central cloud servers. Autonomous networks allow the use of AI accelerators in edge nodes to receive inputs, identify anomalies, and make decisions before this task can be completed in a central control node. Such a distributed intelligence model not only decreases latency but also improves data privacy because sensitive data can be processed locally. More so, with distributed processing, there is sharing of computational resources between the vehicles and the roadside units, enabling efficient use of resources and less traffic congestion in the network. As another example, a car capable of doing processing in real time reaches an intersection with heavy traffic: the car can share some of its ability to compute with a local edge server, which will analyze camera and sensor information in cooperation with the car to decide safe passage. Fail-safe mechanisms are also supported by edge computing because local intelligence can be maintained to guarantee that the cars can still be on the road even at times when their network connection is disrupted or at places where connection is poor.

11.7.3 Network Slicing and Security Customization

A key 6G architecture capability is network slicing, which allows the deployment of numerous virtual networks on top of one physical network—each optimized to suit a particular set of application requirements. Subdividing the network can be done to support different functionalities in the context of automated vehicles such as emergency traffic, infotainment, real-time navigation, and security surveillance programs [27]. This method gives smart systems the ability to run in contained, safe, and efficiency-optimized systems that align to their work priorities. As an example, a special slice can be allocated to security with access to processing and communication resources having the top priority so that the work on threat detection and mitigation is not interrupted. Further, the slices can be dynamically modified depending upon the environmental situation, vehicle behaviors, or the level of threat available, creating adaptive and context-sensitive defense positions. The integration of intelligent systems and programmable network slices is also able to support quality-of-service enforcement and resource availability conditions on mission-critical activities, which eventually results in a thoroughly dynamic and adaptable autonomous transport ecosystem.

11.8 Challenges and Future Research Directions

The process of securing AV networks and intelligent CPSs becomes more complicated as 6G improves the networking and communication capabilities of AVs. Although the combination of AI, ML, edge computing, and blockchain has already been proven to have positive effects on increasing rates of threat detection and ensuring the safety of communications, there are multiple urgent issues to be addressed. These issues are associated not only with potential technical scalability and flexibility of models implemented but also with the overall consequences of changing the threats and realities of the way to a complete implementation of an autonomous and safe transportation network. Such problems are essential to solving and ensuring the promotion of resilient and trustworthy intelligent vehicular ecosystems that can survive in real-world conditions.

11.8.1 Scalability and Adaptability of Detection Models

The scalability and adaptability of threat detection models is one of the greatest obstacles on the path of securing intelligent autonomous systems. Because the connected vehicles, roadside infrastructure units, and other devices fitted with sensors multiply exponentially, traditional models with centralized or fixed servers are insufficient in a 6G context. Detection frameworks are required to operate horizontally on thousands (or millions) of nodes in a real-time fashion. In addition, AVs face extremely variable environments, including the countryside promenade and the busiest corners in the city, which the detection models must adjust contextually. Most existing models tend to not be very generalizable in different situations and need to be retrained frequently, which is computationally prohibitive and inefficient in real-time applications. Future studies should be aimed at creating light distributed AI models in a way that they can be self-adaptive and can adapt to the environment by constantly learning it and implementing changes into the AI model without disruption of the car functions.

11.8.2 Evolving Threats in AI-Driven AV Ecosystems

The emergence of intelligent systems within the AV networks is also finding its reflection in the intentions of attackers who are increasingly targeting exploits that could be applied to AI-based elements. Adversarial ML and data poisoning are just two such attacks that present a serious threat to threat detection mechanism integrity; in the former case, the input is manipulated in subtle ways to fool the model, whereas in the latter, the dataset given during training is corrupted. Moreover, a few years ago, zero-day attacks and AI inversion methods were new avenues of cyber warfare. The threats are especially threatening in real-time CPS environments where a late or wrong response may cause disastrous outcomes. Countering these emerging challenges will require further studies of effective AI methods, including adversarial training, explainable AI, and AI model watermarking, which increase learning-based systems' transparency, traceability, and level of resilience. Also, a partnership among the academic sphere, industry, and regulatory agencies is crucial to future-proof and take care of the distinctive ethical and security concerns of the current AI-driven self-governing heterogeneous systems.

11.8.3 Roadmap to Fully Autonomous Secure Transportation

The accomplishment of a fully autonomous and safe transportation system depends on more than simply technical problems, with regulatory, infrastructural, and societal courses of action also motivating the achievement. The roadmap should focus on the development of common security standards among self-driving vehicles so as to be interoperable across manufacturers and messaging standards. The ability of all types of environments to use the same security policies can be offered by interoperability frameworks together with AI-enabled monitoring of compliance. Furthermore, the further development of edge AI, quantum-resistant cryptography, and FL will be necessary to help real-time and privacy-preserving operations become a reality. At the same time, the success of AVs among the mass audience depends on clarity, interpretability, and adversity testing. The application of the human-in-the-loop model, which incorporates the autonomous decision-making process into human control with additional applications in high-risk or transitional situations, should be a topic of future studies as well. Lastly, preparing digital-twin systems to model current threats and response plans in the real world would allow for proactive engineering of the next generation of intelligent transportation systems—not only automated but possibly secure, extendable, and accountable.

11.9 Conclusion

The fast overlapping of communication standards of 6G communication on automated car networks has transformed the manner in which CPSs are operated, particularly security and intelligently detecting threats. This chapter has touched on the basic architecture and life cycle of intelligent threat detection systems designed around AVs, and it mentioned how important the environment sensors, actuators, communication interfaces, and real-time processing proved to be. By incorporating AI, ML, and DL algorithms, the platform obtains the capacity not only to identify the threats but also to respond to new threats, including spoofing, APTs, and DoS attacks. Moreover, such technologies as FL, blockchain, and quantum-safe cryptography offer better privacy, trust, and forward-looking security procedures. The argument for interoperability between systems, edge computing, and network slicing points to systems that are context-aware and low latency as a means and end to facilitate large-scale and dynamic vehicular networks. Nevertheless, there is an emerging issue of scalability of the model, changing vectors of threats, and standardization of frameworks that are secure. Thus, the challenge of the future should be the development of more adaptive, distributed, and ethically responsible systems to facilitate an autonomous transportation ecosystem that is resilient and reliable. The next step that we are about to take is crucial to achieving the vision of fully autonomous, intelligent, and secure mobility through the cooperation of researchers, industries, and regulatory bodies.

AI disclosure statement: The author confirms that no artificial intelligence (AI)-generated content or AI-assisted writing tools were used in the preparation of this chapter (including summaries and the abstract). All text, analysis, and conclusions are the result of the author's original work.

References

1. Alnaim, A.K. and Alwakeel, A.M., Zero Trust Strategies for Cyber-Physical Systems in 6G Networks. *Mathematics*, 13, 1108, 2025.
2. Tyagi, A.K. and Sreenath, N., Cyber Physical Systems: Analyses, Challenges and Possible Solutions. *Internet Things Cyber-Phys. Syst.*, 1, 22, 2021.
3. Alsulami, A.A., Al-Haija, Q.A., Alturki, B., *et al.*, Security strategy for autonomous vehicle cyber-physical systems using transfer learning. *J. Cloud Comput.*, 12, 181, 2023.

4. Chattopadhyay, A. and Lam, K.-Y., Security of autonomous vehicle as a cyber-physical system. *7th Int. Symp. on Embedded Computing and System Design (ISED)*, India, pp. 1–6, 2017.
5. Kato, T., Fukumoto, N., Sasaki, C., Tagami, A., Nakao, A., Challenges of CPS/IoT Network Architecture in 6G Era. *IEEE Access*, 12, 62804, 2024.
6. Giannaros, A., Karras, A., Theodorakopoulos, L., *et al.*, Autonomous Vehicles: Sophisticated Attacks, Safety Issues, Challenges, Open Topics, Blockchain, and Future Directions. *J. Cybersecur. Priv.*, 3, 493, 2023.
7. Buchta, R., Gkoktsis, G., Heine, F., Kleiner, C., Advanced Persistent Threat Attack Detection Systems: A Review of Approaches, Challenges, and Trends. *Digit. Threats*, 5, 37, 2024.
8. Barua, A. and Al Faruque, M.A., Special Session: Noninvasive Sensor-Spoofing Attacks on Embedded and Cyber-Physical Systems. *IEEE 38th Int. Conf. on Computer Design (ICCD)*, USA, pp. 45–48, 2020.
9. Sumra, I.A., Hasbullah, H.B., AbManan, J., Effects of attackers and attacks on availability requirement in vehicular network: A survey. *Int. Conf. on Computer and Information Sciences (ICCOINS)*, Malaysia, pp. 1–6, 2014.
10. Durlik, I., Miller, T., Kostecka, E., Zwierzewicz, Z., Łobodzińska, A., Cybersecurity in Autonomous Vehicles - Are We Ready for the Challenge? *Electronics*, 13, 2654, 2024.
11. Abdel Hakeem, S.A., Hussein, H.H., Kim, H., Security Requirements and Challenges of 6G Technologies and Applications. *Sensors*, 22, 1969, 2022.
12. Bhide, P., Shetty, D., Mikkili, S., Review on 6G Communication and Its Architecture, Technologies Included, Challenges, Security Challenges and Requirements, Applications, with Respect to AI Domain. *IET Quantum Commun.*, e12114, 2025.
13. Gallego-Madrid, J., Sanchez-Iborra, R., Ortiz, J., Santa, J., The Role of Vehicular Applications in the Design of Future 6G Infrastructures. *ICT Express*, 9, 556, 2023.
14. Rodriguez, E., A Security Services Management Architecture Toward Resilient 6G Wireless and Computing Ecosystems. *IEEE Access*, 12, 98046, 2024.
15. Suomalainen, J., Ahmad, I., Shajan, A., Savunen, T., Cybersecurity for Tactical 6G Networks: Threats, Architecture, and Intelligence. *Future Gener. Comput. Syst.*, 162, 107500, 2025.
16. Korba, A.A., Sebaa, S., Mabrouki, M., Ghamri-Doudane, Y., Benatchba, K., A Life-long Learning Intrusion Detection System for 6G-Enabled IoV. *Int. Wireless Commun. Mobile Comput. Conf. (IWCMC)*, Cyprus, pp. 1773–1778, 2024.
17. Saoud, B., Shayea, I., Yahya, A.E., Shamsan, Z.A., Alhammadi, A., Alawad, M.A., Alkhrijah, Y., Artificial Intelligence, Internet of Things and 6G Methodologies in the Context of Vehicular *Ad-hoc* Networks (VANETs): Survey. *ICT Express*, 10, 959, 2024.

18. Ahammed, T.B., Patgiri, R., Nayak, S., A Vision on the Artificial Intelligence for 6G Communication. *ICT Express*, 9, 197, 2023.
19. Alsubai, S., Umer, M., Innab, N., Shiaeles, S., Nappi, M., Multi-scale Convolutional Autoencoder for Anomaly Detection in 6G Environment. *Comput. Ind. Eng.*, 194, 110396, 2024.
20. Prathiba, S.B., Raja, G., Anbalagan, S., Gurumoorthy, S., Dev, K., A Hybrid Deep Sensor Anomaly Detection for Autonomous Vehicles in 6G-V2X Environment. *IEEE Trans. Netw. Sci. Eng.*, 10, 1246, 2023.
21. Koller, R., *et al.*, Anatomy of a Real-Time Intrusion Prevention System. *Int. Conf. on Autonomic Computing*, USA, pp. 151–160, 2008.
22. Puspitasari, A.A., An, T.T., Alsharif, M.H., Lee, B.M., Emerging Technologies for 6G Communication Networks: Machine Learning Approaches. *Sensors*, 23, 7709, 2023.
23. Yahia, S., Loscri, V., Saxena, P., Enhancing Security in 6G Vehicular Networks: Leveraging VLC and MMW Integration and Cooperative Relaying Technique. *IEEE DASC/PiCom/CBDCom/Cyber.Sci.Tech.*, UAE, pp. 0711, 2023.
24. Barbieri, L., Savazzi, S., Brambilla, M., Nicoli, M., Decentralized Federated Learning for Extended Sensing in 6G Connected Vehicles. *Veh. Commun.*, 33, 100396, 2022.
25. Zhou, Z., Wang, M., Huang, J., Lin, S., Lv, Z., Blockchain in Big Data Security for Intelligent Transportation with 6G. *IEEE Trans. Intell. Transp. Syst.*, 23, 9736, 2022.
26. Kim, M. and Park, J.H., Quantum-Resilient Security for 6G Networks: A Comprehensive Survey on Challenges, Solutions, and Research Opportunities. *J. Supercomput.*, 81, 1086, 2025.
27. Boualouache, A., Jolfaei, A.A., Engel, T., Multi-Process Federated Learning with Stacking for Securing 6G-V2X Network Slicing at Cross-Borders. *IEEE Trans. Intell. Transp. Syst.*, 25, 10941, 2024.

12

Case Study on Cyberphysical Human Systems for Safe Human–Robot Collaboration in a Shared Workplace

H. Nishanthi[1]*, P. Divya Prabha[1], N. Nandhine Shree[1] and Raffaele Mascella[2]

[1]Department of M. Tech Computer Science and Engineering, Sri Ramakrishna Engineering College, Vattamalaipalayam, Coimbatore, Tamil Nadu, India
[2]Department of Communication Sciences, University of Teramo, Teramo, Italy

Abstract
In modern manufacturing, flexibility and advanced mechanization are essential to meet diverse product requirements and the shifting preferences of customers. Under these circumstances, human–robot interaction achieves this by blending the ingenuity and expertise of humans with the accuracy and strength of robotic systems. Cyberphysical human systems (CPHSs) boost collaboration by linking devices such as sensors, actuators, and automated systems to intelligent programs capable of analyzing information instantly and executing operations. Within barrier-free work environments, the CPHS detect individuals' movements and activities, define secure zones, and act autonomously. As an illustration, the systems reduce speed or come to a halt to prevent collisions. This chapter elaborates on the research findings on using multiple sensing devices that allow the avoidance of accidents. According to the research studies, the system can be used to locate, with the use of a number of sensors, more exactly and react more swiftly, therefore enhancing workplace security. Manufacturing systems supported by CPHS are intelligent, adaptable, and user-friendly. Additionally, they foster development by raising awareness and safety.

Keywords: Human–robot interaction, cobot, mobile robot, wearable robot, assistive robot, CPS layered architecture

**Corresponding author*: nishanthi.h@srec.ac.in; ORCID: 0009-0004-9997-3140

Anitha Velu, Prasanth Aruchamy, Raghu Ramamoorthy, Rajesh Kumar Dhanaraj and Manjula Arunraj (eds.) Cyber Physical Human Systems: Emerging Trends and Applications, (271–292) © 2026 Scrivener Publishing LLC

12.1 A Run through of Human–Robot Interaction

In recent years, manufacturing practices have undergone significant changes. Adaptive production approaches, such as mass customization, are gradually replacing older mass production methods [1]. The primary reason for the shift is the necessity for businesses to adapt to a global market that is rapidly changing. These days, manufacturers solve complex and unforeseen issues with the help of both human capabilities and equipment. In such a scenario, human–robot interaction (HRI) becomes highly important [2]. It enables the combination of people and robots to work on the same production lines. Resolving any potential sharing security concerns, nevertheless, becomes a priority as they work together more. Among the most difficult issues in interacting between humans and robots is determining how to balance the safety of workers and still be productive.

Humans are unpredictable; however, robots are accurate yet cannot adapt to new circumstances. The production process can be maximized through combining how fast and precise the robot is compared to human judgment. The problem is that despite their reliability, traditional safety measures can limit the level of production that can be achieved with the help of an automated system. Thus, the most important task is to design innovative safety systems that respond quickly, ensure the safety of workers, and retain productivity levels [3]. Multiple ways of humans and robots cooperating—including sharing a workspace or working on particular tasks together—have been explored by various researchers and scientists.

The findings indicate that teams of humans and robots frequently perform better than either one working alone. Advantages like these will depend on the capability of the safety system to detect and react to possible collision situations. Safety methods are usually divided into two classes: preimpact and postimpact. Preimpact methods prevent accidents through the constant measuring of the distance between the human and the robot, stopping or slowing the robot when the distance becomes unsafe. Postimpact methods operate after contact has occurred by reducing the force to safe levels. Although preimpact methods are generally safer, they can slow production, because they are conservative; the balance of safety and productivity is a continuing challenge in system design.

Various researchers have been working on developing sensor-based systems to improve preimpact collision avoidance. The development of Kinect sensors—for example, optic tracking and force or torque sensors—is designed to detect the presence of a human in proximity and undertake safety actions. Modeling methods have used 3D shapes of humans and robots to predict potential collisions. However, response time once again makes these methods unsuitable. For real-time performance in busy factory settings, where the use of these systems will be most practical, such limitations are highly undesirable. The researchers are investigating an enhanced approach: a cyberphysical system (CPS) based HRI. It integrates sensors, data, communication, and control into one smart, adaptive system. In such a system, physical devices such as robots, sensors, and humans are interconnected, coordinated, and managed with software responding in real-time to real-world conditions.

A robotic CPS should be autonomous, have knowledge about its surroundings, and change its behavior based on real-time human activities and environmental changes to ensure safe collaboration. In existing studies, a typical CPS would include three main parts: the human component, the physical component (PC), and the computational component. Safety is maintained by keeping a changing safety distance between robots and humans. The system operates with a closed-loop controller and a collision detection module to read sensor data while issuing commands to slow it down or redirect it [4]. Such a system would meet real-time requirements with the aid of fast processing and vision systems. To validate it, the researchers designed a CPS prototype for the automotive manufacturing test case where one human and one robot were cooperating using optical sensors and real-time monitoring software.

We were able to assess the system effectiveness by timing how quickly it can recognize a person and respond accordingly. Indeed, the system is faster and more sensitive than a commercial system. This would imply that potentially, the CPS we have suggested may be more secure and effective. In other words, it is not only the compliance of guaranteeing that the cooperative human–robot works in manufacturing industries but also the compliance with the safety regulations. There is a need for smart systems that can understand and respond to the events unfolding on the ground. CPSs are combining physical machines and high-performance computing to make workplaces safer and more productive. Future research will be carried out on the accuracy of the sensors, latency, and the flexibility of the systems to make such systems usable for more industries.

12.2 Challenges of HMI

Protecting human workers working in proximity to automated technologies, whether coworking in concurrent tasks (e.g., concurrent building activities) or coexisting in sideways communication (e.g., being in the same region that has portable robots moving objects), involves something more than physical barriers or slow-moving machinery. New production requires smarter, situation-savvy interventions, which may dynamically estimate and respond to changing environmental situations. Human safety in these scenarios requires the seamless integration of a number of important components. First, shop floor information has to be continuously received and revised to faithfully represent real-world, ever-changing conditions of the workplace. The perception of workers, machines, and transmissible objects, the determination of their eventual behaviors and motions, and real-time, accurate data-obtaining systems are needed to support the maintenance of robot systems and human operators—e.g., cameras, LiDAR (light detection and ranging), ultrasonic, and wearable sensors—to obtain data of robot systems and human operators. Placement of sensors must be deliberate and concurrent for gaining complete situational awareness.

Additionally, with powerful communications, all participants—robots, edge devices, safety controllers, and web-based controls—can depend on fast, low-latency data transfers. In fact, no intelligent sensor or algorithm can prevent an accident without low-latency communication [5, 6]. With this in mind, intelligent software components shall be capable of logical decisions between processing incoming data and speculating on possible dangers. These application modules support the "brain" of the system—managed robot activity—and can dynamically adjust workflows in response to unexpected events or changing safety requirements.

The five primary types of processes include job shop, batch, repetitive, continuous, and project, based on the type of product to be manufactured and the number of products to be manufactured. Each process has its merits and demerits; therefore, one has to consider these processes before making a decision on which one to select as the best. Flexible machines and skilled laborers are required in this process because the work is constantly changing, yet it may be costly and hard to plan. The disadvantages of this type are its inflexibility and low diversity, which are the strengths of its volumes of high efficiency.

The project process type is typically selected in the cases of project work—e.g., filming a movie, publishing a book, or building a dam. The nature of the project type may include the features of each type due to the variety of projects. Therefore, the five types of simplest procedures and their pros and cons must be outlined through compare and contrast. Job shops are sometimes necessitated when there is a need for producing small volumes of unique products; batches are used in medium volumes and work variety. Continuous and repetitive types were developed following the industrial revolution and they are currently the types of high-volume production for the mass consumer.

12.3 Importance of Safe Human–Robot Collaboration

Incorporation of using robots into aspects of human work in the industrial workplace has become one of the most trending ways for an organization to conceptualize a vision of work amongst technologies as industrial workspaces continue to evolve further in this new age of manufacturing. Collaborative robots, or "cobots," are those robots that are designed to work alongside human beings to perform repetitive, challenging, or menial jobs. With the machines having the capability of complementing human ingenuity with precision, there has been a strong jump in productivity, reliability, and efficiency, but just like any other innovation, there are means to take and approach safety effects.

Safety in human–robot collaboration is a technical requirement and a minimum prerequisite for trust, work performance, and long-term sustainability of hybrid human–robot workspaces. Because cobots share the same physical colocation as human workers, unlike isolated, cage-contained, or enclosed industrial robots, their close proximity in working distance implies that any malfunction, communication error, or delay in response may cause severe injury or deadly accidents. Thus, real-time human presence detection, soft working distances, and real-time response of the robot are highly important.

Moreover, all the safety features, such as proximity sensors, vision systems, and control algorithms, provide not only safety but also general system optimality [7]. For instance, robots can potentially vary their speed or motion in relation to human proximity so that the system achieves an optimum trade-off between productivity and safety. Thus, there is no need for the robots to move slowly continuously, yet when necessary to give safety. This responsive action enables the continued workflow with lower risks. Regulators do not treat compliance with the widely accepted international standards of safety, such as ISO 10218 or ISO/TS 15066, as something that is left to choice. Those standards define how to consider safety in design, evaluate threats, and always monitor collaborative systems. The standards protect not just the employees but also the companies from litigation and possible claims arising from accidents. Safely harmonious human–robot collaboration is the essence of clever production.

12.4 Types of Robots Used in Collaborative Setup

In secure boxes, unique robots are used in manners that do not differentiate them from human beings. Robots and humans can be in the same space, sharing tasks with robots completing tasks that require strength, accuracy, or repetition, and humans making decisions or performing tasks that require multimodal dexterity [8, 9]. Robotic technology will look different depending on the robots and the amount of human interaction they can have. Knowing these types is useful when designing safer and more productive working environments.

12.4.1 Collaborative Industrial Robots (Cobots)

Cobots are designed specifically for humans to be in close and direct contact within the same workspace. Cobots, unlike traditional robots, are lightweight, compact, and furnished with sophisticated sensors and force limiters, which will enable them to detect humans and modify their actions. Principle efficiencies of cobots are that they can safely execute repetitive operations, including assembling, packaging, or palletizing, whereas the human employee is focused on aspects of the job that involve evaluation, or tasks requiring and maintaining a degree of flexibility. The majority of prevalent cobots are provided by companies such as Universal Robots (UR), Fuji Automatic Numerical Control (FANUC), and Keller and Knappich Augsburg (KUKA).

12.4.2 Mobile Robots (Autonomous Mobile Robots and Automated Guided Vehicles)

Autonomous mobile robots (AMRs) and automated guided vehicles (AGVs) deliver material within factories and warehouses. AGVs follow magnetic strips or predefined paths, whereas AMRs utilize sensors and onboard intelligence for navigation in dynamic environments. AMRs are needed in collaboration because they save human transportation effort as well as share loading and unloading and potentially share paths. They are used in logistics and just-in-time manufacturing systems.

12.4.3 Dual-Arm Robots

Similar to a human with two hands, dual-arm robots perform complex assembly tasks with concurrent manipulation. Dual-arm robots can manipulate both sides of a product, or work in a common workspace next to a human teammate [10]. Dual-arm robots can be suitable for tasks where precision is required, such as in electronics or automotive manufacturing, because they can simulate a human's physical coordination and fluidity in coordinated movement.

12.4.4 Wearable Robotic Systems (Exoskeletons)

Even though these are not robots, robotic exoskeletons assist the human worker by reducing the human's physical stress from strenuous tasks such as heavy lifting or working overhead. These wearable machines enhance human strength and endurance, reduce injury risk, and improve posture, representing a growing part of human–robot collaborative systems in manufacturing and warehousing.

12.4.5 Assistive Robotic Arms

These are portable, folding robotic arms whose function is to help out in a few operations such as the transportation of tools, positioning of parts, or handling fragile items [11]. Their function is to work alongside the human worker and not to replace the human, whereby they are most useful to use when there are precision industries involved such as assembling medical gadgets or electronics.

12.5 System Architecture of Cyberphysical Human System

The architecture of a cyber physical human system (CPHS) for human–robot cooperation is designed to provide seamless and integrated connectivity between the digital and physical worlds (Figure 12.1). The architecture will be crucial in production settings when human beings coexist in their environment alongside robots to provide safety, flexibility, and productivity. It is based on an interrelated and tiered architecture that accommodates real-time observance, smart decision-making, and control responses based on responsiveness.

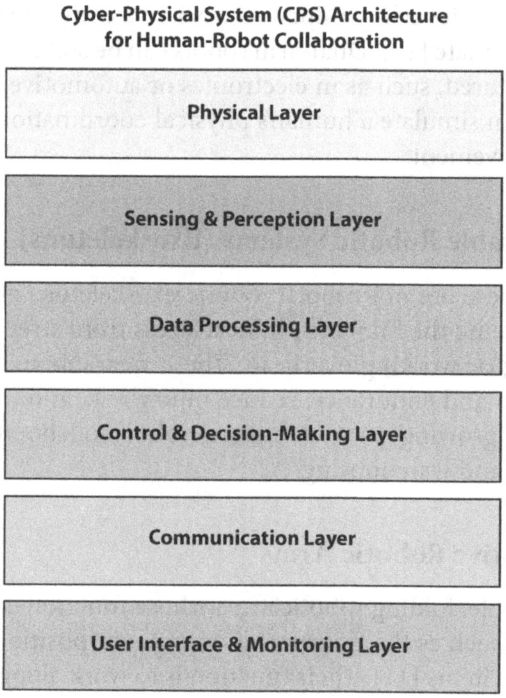

Figure 12.1 System architecture of the cyberphysical human system.

12.5.1 Physical Layer

The physical layer contains all PCs found along the factory floor, such as industrial robots, sensors, human beings, machines, conveyor belts, and mobile robots such as AGVs. All these agents carry out physical activity equivalent to data that are asked of their movement and the environment in which they are present [12]. Robotic-arm positions, employee positions, or AGV paths are some examples of physical data found in this level.

12.5.2 Sensing and Perception Layer

This level utilizes sensors—depth cameras (e.g., Microsoft Kinect), LiDAR, infrared sensors, wearables, etc.—to perceive the real world. Sensors detect the presence of human beings, perceive movement, calculate distance, and perceive environmental parameters. It aims to provide highly dense, real-time data about the work environment and how it varies.

12.5.3 Data Processing and Processing Layer

When the data are captured, they can then enter a layer for data filtering, data analysis, and data interpretation. This layer is where the computational algorithms—such as machine learning (ML) and rule-based systems—route the data to be able to gather useful insights [13]. To illustrate, the system can measure the distance of the arm of the robot in relation to the arm of the human worker, or measure a potentially colliding path. Edge computing or cloud computing can handle big data effectively.

12.5.4 Control and Decision-Making Layer

This is the "brain" of the CPHS. The control and decision-making layer makes decisions and issues control commands based on information received and processed. An example could be if a person walked into the safety zone around an autonomous robot, the control layer may decide to slow the robot, stop the robot, or redirect the mission of the robot. Decision-making could involve ranking jobs, handling exceptions, or changing modes from manually operated to an automated functioning mode.

12.5.5 Communication Layer

Reliable and low-latency communications are important in real-time responsiveness. This layer ensures that all elements—robots, sensors, controllers, and monitoring systems—have continuous communication. This layer uses wired or wireless protocols to exchange data, such as Ethernet, Wi-Fi, Bluetooth, or industrial IoT standards [14]. Also, it ensures the subsystems remain synchronized in real time.

12.5.6 User Interface and Monitoring Layer

The top layer offers the interfaces utilized by human supervisors and operators. Dashboard displays allow the human operator to view the status of the system, indicate alerts, provide a visual representation of the sensed feedback, and allow for manual intervention or override when necessary. This enables the human to observe the system's condition, identify potential risks, and alter its behavior as needed.

12.6 Sensors and Perception Systems for Human Detection

Accurate and timely human presence detection is important for safety reasons in a human–robot collaborative environment where humans and robots work together. It is possible when advanced sensors and sense perception systems are implemented in the CPHS. These technologies are the eyes and ears of the system and the critical information is used to make decisions about when and how to act to avoid collisions and interact in a safe manner.

12.6.1 Depth Cameras and Vision-Based Systems

Depth-sensing cameras such as the Microsoft Kinect, Intel RealSense, or stereo vision systems are some of the more common tools used for human detection [15]. These cameras have the ability to capture red–green–blue (RGB) color images and depth data to provide the system with the opportunity to create a 3D model of the scene. We have completed a human detection system based on detecting skeletal shapes and tracking locations and activities, thereby giving the system the ability to follow where someone is, what they are doing, and how close they are to a robot.

12.6.2 LiDAR

LiDAR sensors measure the distance using laser light and create a high-resolution 3D image of the physical world. These devices perform well with multiple moving targets in dynamic worlds and provide high accuracy, and they are usually implemented on a mobile robot platform for human and obstacle perception along the path.

12.6.3 Ultrasonic and Infrared Sensors

Both ultrasonic and infrared sensors can be used for detecting the presence of humans at close range. Ultrasonic sensors emit sound waves and react to the objects by timing the delay between the emission of the sound and its echo. Infrared sensors react to heat emitted by people or to reflected infrared light. They are much simpler devices than LiDAR or cameras, but inexpensive and typically used with other sensors for redundancy; for the sake of caution, they are often used redundantly.

12.6.4 Use of Wearable Sensors and Tags

In some systems, operators could wear RFID (radiofrequency identification) tags, ultrawideband beacons, or inertial measurement units, and wearable sensors could allow the system to track the operator's position and movement with very accurate fidelity. Although this involves humans being required to wear devices, it does provide strong tracking performance, even in unviewable situations.

12.6.5 Sensor Fusion for Better Perception

No sensor is ideal in every situation. Therefore, sensor-fusion methods are commonly used, integrating information from multiple sensor modalities to construct a more precise and robust representation of the environment. For instance, merging depth cameras with LiDAR can enhance detection quality in bright as well as low-light situations and, furthermore, offset occlusions or sensor noise.

12.6.6 Real-Time Human Tracking Algorithms

The software used to interpret sensor data is important for the operation of brain systems. Machine vision and ML algorithms read images and signals to identify people, gauge distances, track motion, and predict intentions. These algorithms allow robots to anticipate human behavior and respond ahead of time by braking, turning around, or coming to a halt.

12.7 Human–Robot Interaction Models and Strategies

With the increasing intelligence and pervasiveness of automation, human–robot association is shifting rapidly. Instead of functioning in isolation or behind security screens, humans and robots are communicating in the same environment. To make such interactions efficient as well as secure, standardized interaction models and clear-cut strategies are mandated. Those define how human beings and robots interact, coordinate, and react to each other's motion in real time.

12.7.1 Interaction Models

HRI models are models that formulate the relationship between the human and the robot that are involved in task-sharing interaction. The nature of the model is based on the closeness and time of work between them. Under the coexistence model, human beings and robots work near one another without sharing the workspace. Each performs different roles, and physical distance is ensured as it is safe [16]. The sequential collaboration model allows humans and robots to share a task but at different times; for instance, a robot might insert a part, and later, the human operator completes the process by tightening or assembling it. The cooperation model enables both to work on the same task simultaneously, though on different components, requiring continuous observation and coordination between them. Finally, the collaboration model represents the most integrated form of interaction, where human beings and robots collaborate with each other in real time and in the same space. In such an arrangement, the robot must have advanced sensing, communication, and decision-making abilities to coordinate efficiently and safely with the human operator.

12.7.2 Interaction Strategies

By using the interaction strategies, we can use opportunities to set the way the robot takes hints from the human input and dynamic environments to establish a proper and effective human–robot collaboration. The shared autonomy task is characterized and directed by a human and a robot, and there are several approaches to this problem. For example, the human could make fine positioning and the robot could record gross movement. Finally, it is more efficient without compromising on safety.

Smart machines understand in advance what people want to do. With sensors combined with learning software, devices are anticipating hand motions, eye focus, or repeated patterns to be a step ahead. Because of these features, humans and technology work simultaneously without colliding with each other. If a person increases the speed or changes the direction, the machine will adjust the movement and timing to keep in sync with the flexible responses. The device slows down if there is someone around, and thereby prevents colliding. Instead of simply talking, it uses a combination of blinking signals, short sounds, gentle reminders, text displays, or spoken alerts.

These methods help both humans and robots understand each other better and reduce mistakes. Human safety is the most important part of all these strategies. Robots are made to notice when someone enters their working area, slow down when people are nearby, and stop immediately in case of danger. Some systems even consider how safe a person feels when a robot moves close, so humans can feel more comfortable working with them.

12.8 Implementation

To examine the effectiveness, reactivity, and overall feasibility of the CPHS framework for human–robot collaboration safety, we created an alpha iteration of a comprehensive prototype. The prototype served as a proof of concept useful for determining real-life feasibility and facilitating optimal collaboration of the hardware and software elements in an active factory setting. In the middle of this deployment, the cyber components perform the major functions of human and object detection, real-time distance sensing, safety decision-making, and control logic executions. These software modules were implemented using the C# programming language, as it supports excellent object-oriented design, high performance efficiency, and seamless integration with Windows-based sensor libraries and frameworks.

One of the most important operations in the CPHS is computing distances, which is of great importance in identifying the proximity of human operators to robots. This module estimates depth information acquired by one or more 3D sensors within the shared workspace. These sensors continuously make real-time depth measurements that allow the system to use this information in the creation of a dynamic, ever-changing map of the environment. Through analysis of this information, the CPHS will know when a human is breaching a robot's safety threshold and will take all necessary steps to prevent a collision from occurring. Aside from standard depth data, the system is also provided with skeleton tracking data of human workers—information that describes human posture, limb movement, and direction. This allows the system to understand contextually the distinction between safe and unsafe, to-be-avoided interactions—for example, reaching into a region of the workspace or having discordant motion.

Moreover, the real-time input of the system is provided with coordinates of the end effector (the part of a robot that makes contact with objects or tools). With the ability to simultaneously know the human and robot components of the system, the CPHS is able to execute complex collision prediction and execute reactive or proactive measures to resolve the situation, such as slowing down or stopping the robot motion. To help with high performance and precision, the system uses custom Kinect SDK libraries, which have powerful skeletal tracking abilities, as well as RGB-D sensor support.

These specialized libraries were in turn further refined to realize high performance in skeletal recognition and reducing processing latency such that the prototype may have near-real-time response rates, which is an essential feature in safety-critical applications. The prototype development is a great milestone in production-viable applications of CPHS solutions. This shows the potential of smart, sensor-based systems to create safer environments with open, adaptive human–robotic interactions across any barriers. In proving the possibility of the unification of high-end sensing, spatial data computation, and real-time control within a single thread, it opens the door to additional future-enabling capabilities such as artificial intelligence (AI)–supported prediction of behavior, scalability to large workspaces, and compatibility with industrial safety norms.

12.9 Risk Assessment and Hazard Mitigation Techniques

As robots continue to join the workforce alongside working employees, safety is vital. Therefore, high-quality risk assessment and hazard control processes must be in place so that an accident does not happen and work continues. The processes will allow for the identification of potential hazards, evaluation of the severity of risk involved in different situations, and implementation of appropriate measures to eliminate and reduce risk.

12.9.1 Understanding Risk Assessment in Human–Robot Collaboration

Risk assessment is a systematic process that is utilized to evaluate potential hazards in a human–robot cooperation setup. It encompasses the identification of sources of harm, analysis of the probability and severity of incidents, and determination of acceptable levels of risk. The process is founded on global safety standards such as ISO 12100 (machinery safety) and ISO/TS 15066 (safety of cobots). In a human–robot cooperation scenario, typical hazards are unintended robot motion, crashes during hand operation, system failures (sensing malfunction, software glitch), and environmental conditions (inadequate lighting, over-cluttered workspace).

12.9.2 Significant Steps in Risk Assessment

A good risk evaluation has a number of steps that are critical in assuring safety within a human–robot workspace. The first stage is hazard identification, which means the identification of potential sources of harmful factors, which may include sharp tools handled by robots, sudden movements of the robot, or even unsafe working conditions. The next step after the detection of these hazards is risk estimation, in which the probability of each incident and the potential impact are thoroughly researched. Thereafter, risk assessment is implemented to determine the degree of risk tolerance or whether extra safety measures are necessary. Lastly, risk reduction emphasizes acts that should be taken to reduce the identified risks. It might involve redesigning the system, providing barriers or sensors, and applying behavioral control interventions to ensure the working environment is safer not only to humans but also to robots.

12.9.3 Hazard Mitigation Strategies

There is risk assessment, and then safety measures are planned and implemented to mitigate or eradicate the risk hazards identified. Protective measures are one of the ways that entail the installation of physical protective measures, safety fences, light curtains, pressure mats, or interlock devices that automatically shut down the robot when an individual is in a restricted area. Speed and separation monitoring (SSM) is another strategy in which the robot increases and decreases speed depending on the distance of human operators, at a full speed when no one is nearby and in deceleration or stopping when a person is nearby.

The power and force limitation (PFL) technique makes sure that the robots operate within the range of safe forces such that even in the event that they accidentally collide with a human being, no harm is caused. Besides this, an emergency stop system is added, and easy-to-access buttons are provided, which immediately halt robot movement whenever danger is detected. Redundant detection systems implement more than one sensor—cameras, LiDAR, or wearable tags—to guarantee a better human detection system and a reduced probability of failure of the system. Lastly, dynamic workspace surveillance involves vision systems, which operate in real-time tracking to observe the immediate surrounding and take prompt action in case of any undesirable occurrence.

12.9.4 Continuous Safety Validation and Improvement

Too often, monitoring and safety verification are not sustained following the completion of mitigation measures. Controlled experiments and simulations validate that the safety systems work as intended. The human operator input must also be evaluated to maximize processes and benefit safety culture.

12.10 Machine Learning and AI Integration for Dynamic Decision-Making

ML and AI are reshaping human–robot collaboration systems and the way decisions get made in adaptable, real-time scenarios. There was a classical system of robots that was founded on programmed rules and independent behavior schemes. They were designed in such a way that they discouraged response and adaptability. In contrast, AI-enabled the CPHS to make information-driven, context-body decisions using real-time data, learning algorithms, and predictive analysis to optimize both safety and productivity at the shop floor.

12.10.1 The Function of Artificial Intelligence in Collaborative Robotics

Robots have the capability of managing complex scenarios and acting intelligently in response to environmental remedies through AI. AI also offers robots with a number of capabilities that make them safer and more efficient to collaborate with humans. Robots have the ability to monitor human behavior through studying movement patterns and can determine what that individual will do next. They also have the ability to master task sequences and modify their activities according to workflow within a production facility.

Moreover, AI-powered robots can recognize humans or objects through the computer vision and gesture recognition system; thus, they learn human signals more efficiently. They will also be able to change their speed, force, or direction according to the human movement and distance, which assists in the prevention of accidents and the facilitation of smooth collaboration between humans and robots. Spontaneous decision-making and spontaneous action must be possible to minimize threats of accidents, simplify the work process, and establish as natural an interaction as possible between humans and robots in a changing environment and human activities in real time.

12.10.2 Machine Learning for Ease of Development and Learning in Work Cells

ML enables robots to learn and gain more knowledge during their work experience. Various kinds of algorithms, which include supervised, unsupervised, and reinforcement learning (RL), assist robots to adapt to dynamic conditions in industries. During the process of supervised learning, robots are made to understand how to perceive when humans are nearby and categorize their actions as either safe or hazardous. RL assists robots to enhance their behavior and paths by experimenting and staying within the safety boundaries. Unsupervised learning is used to identify the anomalous sensor data patterns, which is used to indicate a faulty system or abnormal behavior. These ML models upgrade themselves in the long term, thereby improving the system and becoming more accurate and efficient to the changing work environment.

12.10.3 Real-Time Data Processing and Decision Support

AI systems are now able to process vast data in real time with the expansion of edge computing and the advancement of high-performance sensors. This dynamic information aids the robots in making swift and intelligent decisions in the course of operation. AI is able to avoid collisions by monitoring human movement and altering the route of the robot to prevent collisions. It also schedules dynamically, based on the condition of the operator or the alteration of the surrounding environment. Further, AI aids in automatic error correction through identifying instances where the behaviors of a system do not follow the common pattern. Comprehensively, AI-driven decision-making is both reactive and foresighted, which means that robots will be able to act not only once something is done but also to predict and avoid the incidents before they happen.

12.11 Challenges and Lessons Learned from Implementation

Safe human–robot collaboration through the CPHS implementation is a complex project that combines different engineering areas, and these areas are robotics, software, sensor integration, and safety systems. Despite the obvious benefits of these systems—i.e., the improvement of the safety levels, an increase in the number of products, and flexibility—the implementation process is usually full of technical, operational, and organizational challenges. These challenges need to be internalized and the actual deployments can be learned from to improve subsequent implementations.

12.11.1 Sensor Limitations and Reliability of Data

The key problem was the solution to ensure proper and accurate human detection by sensors. Kinect or LiDAR or other vision-based systems are prone to occlusions, light, and reflective surfaces, which could result in a false alarm or failure to detect. It was discovered that multisensor fusion involves a combination of depth cameras and infrared or ultrasonic sensors that increases reliability but also adds complexity to calibration and synchronization. A redundant, multilayered sensor architecture enhances safety but needs to be rigorously tested under a variety of environmental conditions to prevent blind spots and sensor conflicts.

12.11.2 Real-Time Constraints

Building real-time capability for collision detection and robot response is a serious challenge. Depth data processing, safety distance calculations, and issuing robot slowdown or stop commands have to be done within milliseconds to support human safety. Early prototypes of the system took several milliseconds due to computational lag. Offloading processing to edge devices and algorithm optimization for performance (e.g., efficient use of C# code and GPU acceleration) are crucial to address very stringent real-time requirements.

12.11.3 Integration with Legacy Systems

Robots in industrial environments usually have preexisting programmable logic controllers (PLCs) and industrial networks. The integration of the new CPHS in combination with these legacy systems required custom interfaces and middleware that limited deployment speed and made the system more complex. If compatibility should be more with the wider use of accepted protocols (e.g., OPC-UA), integration with installed infrastructure will be much easier.

12.11.4 Human Factors and Operator Trust

Early user testing identified a few operators who were reluctant to work alongside robots even when safety shields were engaged. Human psychology is a significant factor in acceptance [17, 18]. False starts or detection failures were raising concerns among the operators. Transparent communication, training of operators, and user-friendly interfaces provide credibility for collaborative systems. The visual and auditory feedback generated by the system informs operators of the state of the system and motives, making them feel safer.

12.11.5 Complexity of Risk Assessment

Risk assessment that addresses all the possibilities of a dynamic work setting was challenging to complete. Safety requirements are changing; yet, unless there is a certain amount of comprehension of the domain, it is challenging to decipher those safety requirements into system behavior. There should be constant interaction and recurring risks among safety authorities, tests have to be evaluated, and edge-case situations need to be modeled. Formally verified certified safety logic tools could aid as compliance advisors.

12.11.6 System Scalability

The CPHS must be able to scale up continuously because the robots or workstations are incrementally being added without performance being lost. It took an exponential number of sensors; adding robots or human operators added complications to the system. The design of a distributed and modular architecture right from the beginning supports improved scalability. Microservices and edge-based processing support decision-making at the local level while preserving global system cohesion.

12.12 Test Cases of CPHS with HMI

The general-purpose fielded CPHS was deeply tested through a highly detailed test case study under a controlled experimental laboratory machine shop, duplicating actual-world conditions encountered in the automobile industry. The general purpose of this testbed exercise was to test the responsiveness and performance of the software-based control system, specifically for human–robot collaboration. This was focused on protecting the human operator through enforcing and testing numerous adaptive safety policies so that the robot dynamically adjusts its rate of operation, depending on how close the human operator comes. Because the human comes close to the robot—particularly when working in close proximity, like in screwing or assembly—the robot should have the capability to detect infringement of its workspace and quickly trigger protective safeguarding features.

In this specific case, the setting was a robot arm operating from a workstation where a human operator approached the robot to perform a screwing task on an object within close proximity of the robot work envelope. Neither the robot system nor the operator was fenced in, similar to an open, cooperative workspace in high-tech manufacturing cells. As a human entered the proximity region near the robot, sensor-based tracking modules within the system tracked the human's motion, taking skeletal data and depth to report precisely where they were in space, as well as the path of motion, through an ongoing monitoring process. Based on this real-time data, control software would then match the relative separation between the end effectors of the robot and operator to preprogrammed safety limits.

Whenever the operator happened to be in some range of the robot, a safety policy would automatically start, such as slowing down the robot or even completely shutting it down, depending on the level of the proximity threat. Safety measures were in place to try to reduce the chances of an accidental contact while retaining decent levels of productivity. When the operator was moving away, the robot resumed the normal speed, which is where the adaptability of the CPHS comes in. The test also allowed for different response thresholds and means of reaction to be compared so that the best approach for a safety and practical balance could be established.

It was also favored by having an in-lab testbed to test how various iterations and sensor configurations—including a single depth sensor or two depth sensors—would influence the detection rate and latency. It was found that the CPHS could offer rapid response times to humans and low latency to be safely utilized in real time. It highlights the practical relevance of the system, which was created in the context of a car assembly system where robots and human workers would be forced to work side by side and perform scheduled procedures. As a result, the CPHS is a practical and handy tool in dynamic safety control imposition and a valid foundation toward its eventual implementation in a full-scale production industrial environment.

12.13 Conclusion

In HRI, this chapter discusses the role played by the CPHS in ensuring safety and efficiency in state-of-the-art manufacturing. It has highlighted how the synergistic application of smart sensing technology and real-time control systems can enable both robots and humans to share the same working environment without the need to have a physical dividing line and with no loss of safety. This chapter emphasizes that the quick and proper identification of human beings with the assistance of configurations of redundant sensors is essential in minimizing the response time and preventing hazardous situations. CPHS enhances not only safety at the workplace but also flexibility and efficiency in the process of complex work during assembly through the promotion of cooperative presence systems. Predictive models and AI will become more and more part of each other, and this fact will lead to the consolidation of the systems to make decisions more proactive and preemptive, paving the way to really responsive and sustainable interaction between people and robots in the industry.

AI disclosure statement: The author confirms that no artificial intelligence (AI)-generated content or AI-assisted writing tools were used in the preparation of this chapter (including summaries and the abstract). All text, analysis, and conclusions are the result of the author's original work.

References

1. Allen, R.J., Baddeley, A.D., Hitch, G.J., Is the binding of visual features in working memory resource-demanding? *J. Exp. Psychol. Gen.*, 135, 298, 2006.
2. Choi, H.S., Han, C.S., Lee, K.Y., Lee, S.H., Development of hybrid robot for construction works with pneumatic actuator. *Autom. Constr.*, 14, 452, 2005.
3. Halme, R.J., Lanz, M., Kämäräinen, J., Pieters, R., Latokartano, J., Hietanen, A., Review of vision-based safety systems for human-robot collaboration. *Procedia CIRP*, 72, 116, 2018.
4. Kulić, D. and Croft, E., Pre-collision safety strategies for human-robot interaction. *Auton. Rob.*, 22, 149, 2007.
5. Liu, Y., Habibnezhad, M., Jebelli, H., Brain-computer interface for hands-free teleoperation of construction robots. *Autom. Constr.*, 123, 103523, 2021.
6. Liu, Y., Habibnezhad, M., Jebelli, H., Brainwave-driven human-robot collaboration in construction. *Autom. Constr.*, 124, 103556, 2021.
7. Szeliski, R., *Computer Vision*, Texts in Computer Science, Springer, London, 2011.
8. Liu, Y., Habibnezhad, M., Shayesteh, S., Jebelli, H., Lee, S., Paving the way for future EEG studies in construction: Dependent component analysis for automatic ocular artifact removal from brainwave signals. *J. Constr. Eng. Manag.*, 147, 04021087, 2021.

9. Lotte, F., Congedo, M., Lécuyer, A., Arnaldi, B., A review of classification algorithms for EEG-based brain-computer interfaces. *J. Neural Eng.*, 4, 13, 2007.
10. Suzuki, S. and Be, K.A., Topological structural analysis of digitized binary images by border following. *Comput. Graph. Image Process.*, 30, 32, 1985.
11. Zhu, J. and Hastie, T., Kernel logistic regression and the import vector machine. *J. Comput. Graph. Stat.*, 14, 185, 2005.
12. Bock, T. and Linner, T., *Construction Robots*, Cambridge, pp. 14–290, University Press, 2016.
13. Kartika, I. and Mohamed, S.S., Frame differencing with post-processing techniques for moving object detection in outdoor environment, in: *Proc. of IEEE 7th Int. Colloq. on Sig. Proc. and App.*, 7, 172, 2011.
14. Stamou, G.N. and Pitas, I., Object tracking based on morphological elastic graph matching, in: *Proc. of IEEE Int. Conf. on Img. Proc.*, 3, 709, 2005.
15. Karpenko, A., Jacobs, D., Baek, J., Levoy, M., Digital video stabilization and rolling shutter correction using gyroscopes. *Stan. Tech. Rep. CTSR*, 7, 3, 2011.
16. Habibnezhad, M., Shayesteh, S., Liu, Y., Fardhosseini, M.S., The architecture of an intelligent digital twin for a cyber-physical route-finding system in smart cities, in: *8th Int. Conf. on Constr. Engg. and Proj. Man.*, 8, 112, 2020.
17. Jebelli, H., Hwang, S., Lee, S., EEG signal-processing framework to obtain high-quality brain waves from an off-the-shelf wearable EEG device. *J. Comput. Civ. Eng.*, 32, 12, 2018.
18. Jebelli, H., Hwang, S., Lee, S.H., EEG-based workers stress recognition at construction sites. *Autom. Constr.*, 93, 315, 2018.

13

Medical Cyberphysical Systems: A Solution to Smart Health Using Flexible Sensors and Devices

Prema S.[1,2*] **and G. Krishnapriya**[1]

[1]*Department of Computer Science, School of Engineering and Technology, Pondicherry University, Puducherry, Puducherry, India*
[2]*School of Computing, Vel Tech Rangarajan Dr. Sagunthala R&D Institute of Science and Technology, Chennai, Tamil Nadu, India*

Abstract

Traditional hospital-based healthcare systems relied on in-person doctor appointments, posing challenges for the elderly with mobility limitations. With the growing aging population, there is an increasing need for smart, accessible healthcare solutions. In this chapter, an integrated approach to elderly care through the advancement of medical cyberphysical systems (MCPSs), incorporating explainable artificial intelligence (XAI), digital twins (DTs), Society 5.0 principles, and intrusion detection systems (IDSs), is discussed. MCPSs enable continuous, real-time remote monitoring using wearable sensors and intelligent technologies. XAI promotes transparency in artificial intelligence–driven decisions, allowing clinicians and patients to understand and trust system recommendations. DTs simulate patient health in real time, enabling early diagnosis and personalized treatments. Aligned with the human-centric goals of Society 5.0, MCPSs connect the physical and digital realms to deliver adaptive and personalized care. To ensure security, IDSs at the network edge detect threats in real time and protect sensitive health data, forming a secure, scalable, and efficient framework for elderly healthcare.

Keywords: Medical cyber-physical systems, elderly healthcare, explainable artificial intelligence, digital twin, society 5.0, wearable sensors, personalized healthcare, data security

**Corresponding author*: sprema@veltech.edu.in; ORCID: 0009-0003-5752-7516

Anitha Velu, Prasanth Aruchamy, Raghu Ramamoorthy, Rajesh Kumar Dhanaraj and Manjula Arunraj (eds.) Cyber Physical Human Systems: Emerging Trends and Applications, (293–308) © 2026 Scrivener Publishing LLC

13.1 Introduction

The increasing elderly population worldwide presents significant challenges to conventional hospital-centered healthcare models that typically rely on face-to-face doctor consultations. Mobility constraints among older adults often hinder the ability to receive timely medical attention. To overcome these barriers, there is a pressing demand for smart, continuous, and accessible healthcare systems that extend beyond traditional clinical environments. A revolutionary approach to this problem can be provided through the concept of medical cyberphysical systems (MCPSs) that combines state-of-the-art technologies, including wearable sensors, continuous health monitoring, and intelligent data analytics, to offer customized healthcare services. Through the smooth integration of the physical and cyber environments, MCPSs are able to bring about proper communication and coordination in the provision of patient care.

By adopting explainable artificial intelligence (XAI) [1–4], transparency of artificial intelligence (AI)–based decisions in healthcare promotes confidence in healthcare professionals and the patients. Digital twins (DTs) [5] add a new dimension to the system as they reflect the conditions of the patients virtually, assisting in early diagnosis and a personalized approach to therapy. MCPSs, as one of the initiatives inspired by the Society 5.0 vision, which focuses on balancing technological progress and the well-being of people in society, support proactive and adaptive care of the elderly. With the growing interconnected nature of the healthcare infrastructure, security of patient information and integrity of the systems become critical. Application of intrusion detection systems (IDSs) on the network edge can also facilitate detection and prevention of cyber threats, thus guaranteeing a secure network.

This chapter focuses on overall elderly care governance by integrating MCPSs with XAI, DTs, Society 5.0, and IDSs, so as to establish a resilient, secure, and people-based smart health system toward improvement of aged health.

13.2 Background on MCPSs and Smart Health for the Elderly

In the traditional healthcare system, patients who suffer from disease should visit the hospital directly. Once they obtain the appointment, they are allowed to discuss the health and take the clinical decision. Once cured from disease, patients fail to continuously monitor their health due to busy work-life schedules. In case of an emergency, if a doctor is in need of a patient's health record, it will take time to manually search for previous health conditions. To address all these issues, smart health can be implemented to elderly patients' healthcare through the help of sensors and wearable devices that will provide continuous monitoring. Health vital signs will be collected and stored in the server/cloud. When data are stored, there is an ethical concern of data privacy. MCPSs bridge physical health devices with cyber, allowing intelligent health solutions. They integrate wearable flexible sensors, Internet of Things (IoT) devices, cloud computing, AI, machine learning (ML) [6] analytics for real-time monitoring of health.

13.2.1 Major Elements of an MCPSs for Intelligent Health

- **Flexible sensors**: Portable biosensors [7] capturing actual physiological parameters (e.g., heart rate, SpO_2, blood pressure, electrocardiogram).
- **Wireless communication**: IoT-enabled connectivity to have smooth data flow between sensors, mobile apps, and cloud environments.
- **Edge computing**: On-site processing of sensor [7] information to minimize delay and facilitate real-time decision-making.
- **Cloud analytics**: Predictive models powered by AI that crunch health trends and offer early warning signals.

By integrating deep learning (DL) models, particularly explainable deep neural networks, MCPSs enhance the intelligence of these systems, allowing early disease prediction, anomaly detection, and personalized treatment recommendations for elderly patients.

13.3 System Architecture of Smart Medical Cyberphysical Systems Framework

A smart MCPS framework integrates multiple components that work together to ensure real-time monitoring and intelligent decision-making.

13.3.1 System Architecture Workflow

Figure 13.1 shows the architecture of the MCPS framework where the real-time health vital parameters can be read through sensors and transferred *via* the communication layer. Wireless protocols such as Bluetocth low energy, NB-IoT, and ZigBee can be used to transfer data. The edge/cloud computing layer stores the data acquired into the cloud/server.

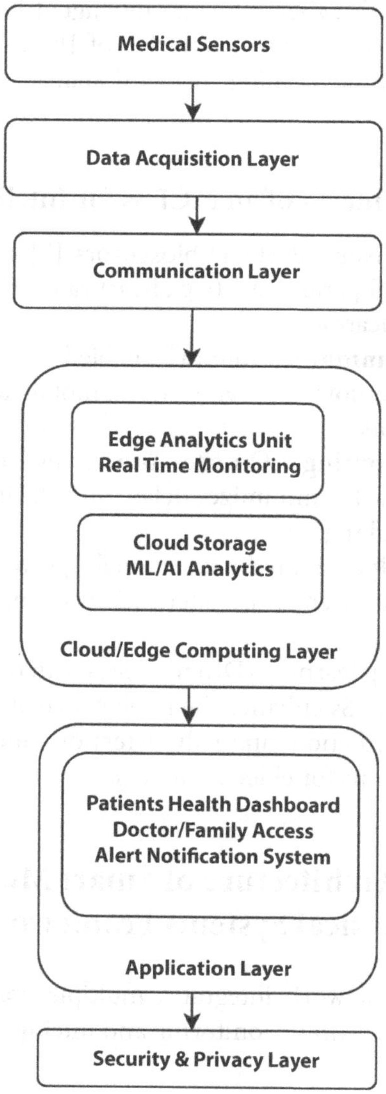

Figure 13.1 Architecture of smart medical cyberphysical systems framework.

ML models shall be used for edge computing in continuous health tracking of a patient's health. The security [8] layer ensures access control, data encryption, and Health Insurance Portability and Accountability Act/General Data Protection Regulation compliance.

If an elderly patient's heart rate exceeds a critical threshold, an automatic alert is triggered, allowing caregivers to intervene immediately. Because health data are sensitive, implementing a strong security layer is essential to safeguard the data from unauthorized access. Following are the various MCPSs discussed that ensure security, trustworthiness, and transparency for enhancing elderly care.

13.4 Smart Health Solutions Using MCPSs and Emerging Trends

13.4.1 XAI in MCPSs

Figure 13.2 shows XAI incorporated with MCPSs to enhance trustworthiness and transparency. The model's interpretability is achieved when the results of the framework evolution target feature are in sync with the results of reasoning and are deemed to be of considerable reliability. However, the utilization of these technologies introduces novel challenges, notably the imperative to safeguard the security and privacy of patients' health-related data within the Internet of Medical Things (IoMT).

As a result, the identification of attacks becomes an essential part of security mechanisms for MCPSs. The attack detection device and state estimation mechanism are modeled under the assumption that a subset of the IoMT sensors is protected. An attack detector can be developed to increase the security of IoMT networks through the concept of sparse observability and a linear relationship among the sensor measurements. A system is called *s*-sparse-observable with respect to any s sensors, if the system is immune to any attacks with *s* sensors. Additionally, when the number of protected sensors $r \geq m$, any sensor *i* with measurements that can be represented as a weighted sum of other sensor readings,

$$y_i(t)=k_{i1}y_1(t)+\cdots+k_{im}y_m(t) \tag{13.1}$$

is likely compromised. Therefore, the maximum number of IoMT sensors that must be secured is *m*. Linear dependencies between sensor data can help detect and locate attacks. Assuming $r \leq m$ and that the matrix *P* contains *r* linearly independent rows, the observer can be set up using a system

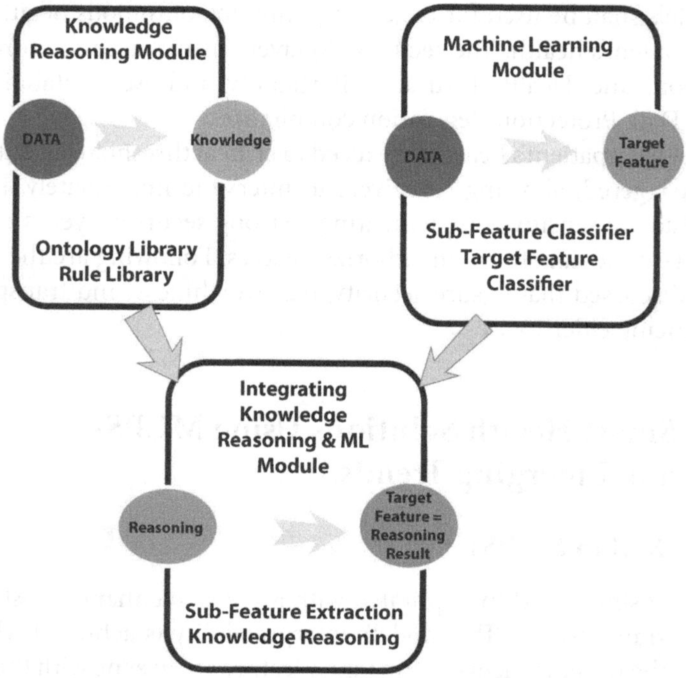

Figure 13.2 Explainable artificial intelligence (XAI) in MCPSs.

of matrices (M, N, L) and a state-space observer model. The observer uses actual sensor data and estimated outputs to calculate a residual signal $r(t)$. If the observer matrix L is chosen so that the matrix $M - L * P\_\{1,\ldots,m\}$ is strictly stable, meaning all eigenvalues lie within the unit circle, then a vanishing residual $r(t) \rightarrow 0$ shows no attack on the m core sensors. Otherwise, if $r(t)$ does not converge to zero, it suggests a compromise.

The parameter initialization, setting the rank of P to m, and the consistency of measurement data form the beginning stage of the attack detection. In case of inconsistencies, an attack is set off. When r is less than m, the rest of the m minus r rows are chosen in the matrix C to derive a largest possible set of linearly independent sensor readings in P to make additional checks and identifications of sensors that are being compromised. It is a potent means of securing the IoMT systems against cyber threats and has the advantage of maintaining the integrity of data about patient monitoring.

13.4.2 Digital Twins

Increased demand for accurate and customized healthcare solutions has drastically grown over the past few years due to the quest to have superior medical outcomes. Besides the original use in manufacturing and car-related industries, the use of DTs is a relatively new field that is bound to make a major impact in the sphere of healthcare. DTs accelerate a more custom experience in diagnosis, treatment, and monitoring as they create

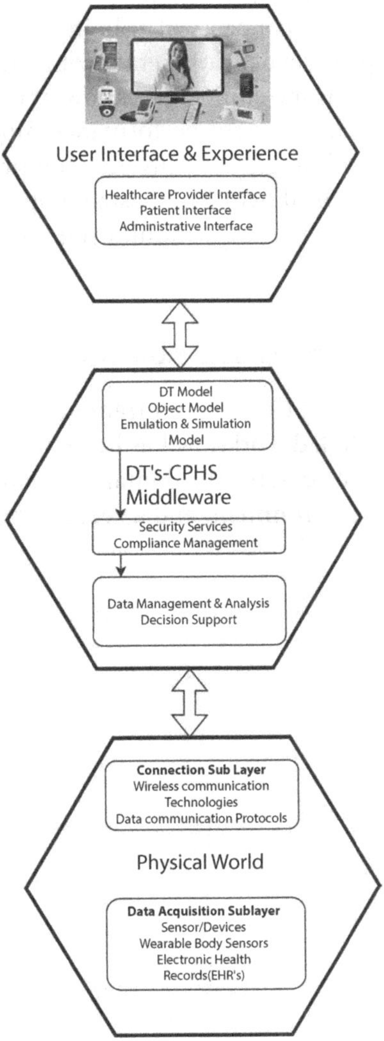

Figure 13.3 Architecture of digital twins in healthcare.

real-time digital representations of the health of a patient. The complexity of modern healthcare requires the healthcare practitioner to rely primarily on his or her experience when making critical decisions, though technological advancements have provided an extension to this so far. This highlights the continued importance of professional development.

Even though integrated healthcare support systems and personalized treatments are still in their infancy, the potential to improve patient outcomes is becoming more widely acknowledged. Figure 13.3 describes the fundamental design specifications and examines the operational and technical difficulties in creating digital healthcare infrastructures. Figure 13.3 illustrates the fundamental concepts of DTs in the medical field, specifically the functions in product development through the user interface layer through wearable [9] sensors, clinical workflow optimization through the DT's middleware layer, and system performance monitoring through the physical world layer. Additionally, it delivers a comprehensive overview of developments in the medical field, emphasizing operational effectiveness, personalized medicine for real-time monitoring [10], and healthcare data management.

13.4.3 CPS-Driven Transformation Toward Society 5.0

Three main layers make up the CPS 5.0 framework [11, 12], as shown in Figure 13.4, and the detailed workflow has been depicted in Figure 13.5 as follows: the cyber layer represents the digital world; the physical layer represents the real-world environment; and a security layer in between those

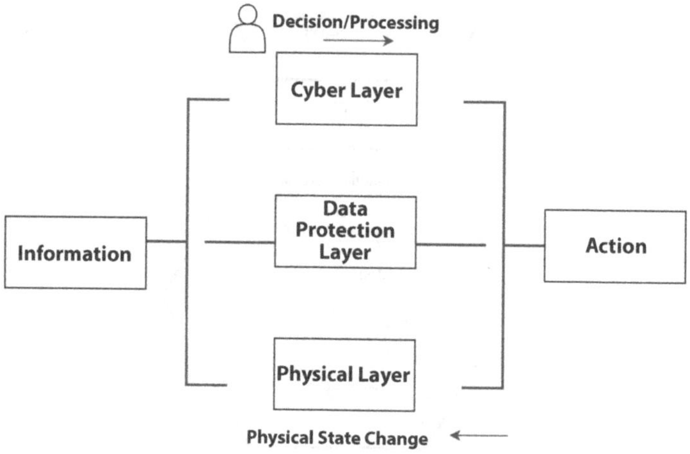

Figure 13.4 Architecture for cyber-driven transformation 5.0 framework.

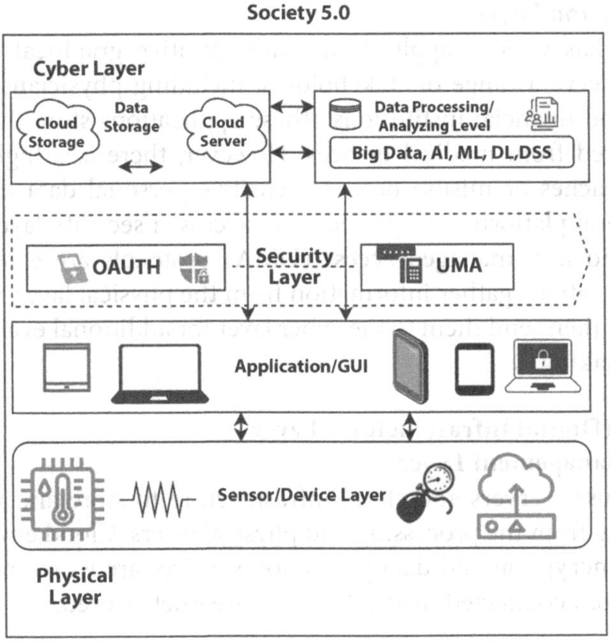

Figure 13.5 Workflow of cyberphysical system 5.0 framework.

two layers guarantees safe interaction. The sensor/device layer, which gathers data, and the application layer, which controls interactions, are the two sublayers that make up the physical layer. Similar to this, the cyber layer is made up of two sublayers: the processing layer, which analyzes data, and the storage layer, which stores data.

(i) Physical Layer

(a) Device Layer and Sensor Layer

Respiratory monitors, sleep apnea therapy devices, sleep diagnostic tools, anesthetic machines, spirometers, dialysis units, infusion pumps, oxygen concentrators, blood glucose monitors, pulse oximeters, and other medical devices, sensors, and actuators are all part of this layer. These gadgets are usually linked to a computer or smartphone, creating a closed-loop system that is less vulnerable to cyberattacks unless it is connected to the Internet through routers or other networked devices. Direct patient interaction with these devices results in the transmission of gathered data to the application layer. Direct and offline communication technologies such as Wi-Fi, Bluetooth, and other smartphone-compatible wired and wireless connections can be used to transfer data.

(b) Application Layer

This layer has various applications with intuitive graphical user interfaces that serve a range of stakeholders, including physicians, hospitals, patients, and financial institutions. These applications store the information gathered from medical devices. However, there is a higher chance of data breaches or misuse because sensitive personal data are handled across several platforms. To protect data access, a security layer is implemented, and user-managed access (UMA) protocols are essential. This layer's applications gather information from the physical layer, show them locally, and then send them to the cyber layer for additional evaluation and decision-making.

(ii) Cyber (Digital Infrastructure) Layer

(a) Data Management Layer

This layer uses servers and cloud infrastructure to store and manage all data coming from the processing and physical layers. Cloud environments that offer encryption and data protection services are where patient data gathered from connected medical devices are safely stored.

(b) Data Processing/Analysis Layer

In this layer, the collected data are analyzed using different algorithms and methodologies. Decision support systems have become more accurate and efficient as a result of developments in AI, DL, and ML [13], as well as the availability of large-scale clinical data (big data). This has helped to ensure the best possible care for patients. Reports are created from the results of this processing and returned to the data storage layer. These reports can be used by physicians to conduct consultations. Given the high volume of personal information and application usage in MCPSs, a robust security layer is crucial to ensure secure authentication and access control.

(iii) Security Layer

This layer implements a ciphertext-policy attribute-based encryption system to preserve the data of the patient. Although privacy is enhanced, this method involves multiple servers that increase the cost of the system, and also there is a problem of trust. Determining which data items require an increased level of confidentiality and providing the appropriate security techniques is the primary purpose of this work.

The security layer of the Society 5.0 framework helps manage the secure sharing of personal information through self-sovereign identity (SSI), UMA, and OAuth technologies. Based on the data protection laws of different countries, it can be implemented differently. An authorization (not authentication) standard—the OAuth 2.0—has become a favorite that grants access tokens to other apps over the user data. UMA operates on a centralized dashboard, based on OAuth 2.0, and enables the user to regulate what users they make available to, when they become available, and how long they remain available. Individual and organizational SSI is portable, verifiable, and lets one access services safely but privately. As an example, one can give access using OAuth and UMA protocols in the situation when a physician has to view the medical history of a patient or his or her diagnostic results. Similarly, payment applications can be used to process the cost of consultation as agreed between both the physicians and patients with relevant information.

To manage these procedures, a specialized security server can be set up in Society 5.0. Tools such as Jakarta EE, Auth0, a12n-server, or Curity Identity Server can be used to implement OAuth, and platforms such as ForgeRock Identity Gateway, Gravitee.io API Platform, or Gluu Gateway can be used to layer UMA. Tools such as Automated Validation of Internet Security Protocols and Applications (AVISPA) can be used to validate these implementations.

This improved framework resolves unresolved data privacy issues and is based on risk analysis and requirement engineering from Society 4.0 CPS. OAuth 2.0 and UMA can be seamlessly integrated without interfering with the current infrastructure by adding a unique security layer that is distinct from the physical and cyber layers. This successfully reduces the risks to data privacy. However, designing and deploying MCPS systems still face several challenges, which are explored in the following section.

13.4.4 Intrusion Detection System

(i) Network Model

Figure 13.6 shows the network model for protecting mobile healthcare systems. As a complete security solution intended to reduce a range of cyberthreats, this architecture incorporates an IDS based on ML. The following essential elements are included in the model:

(a) Healthcare Agents

These include end users such as patients with sensors [14] and smart medical devices, as well as service providers, such as hospitals, pharmacies, and insurance companies. Each agent connects to the system *via* an edge connectivity layer, where all healthcare-related data are transferred to nearby mobile-edge devices [15].

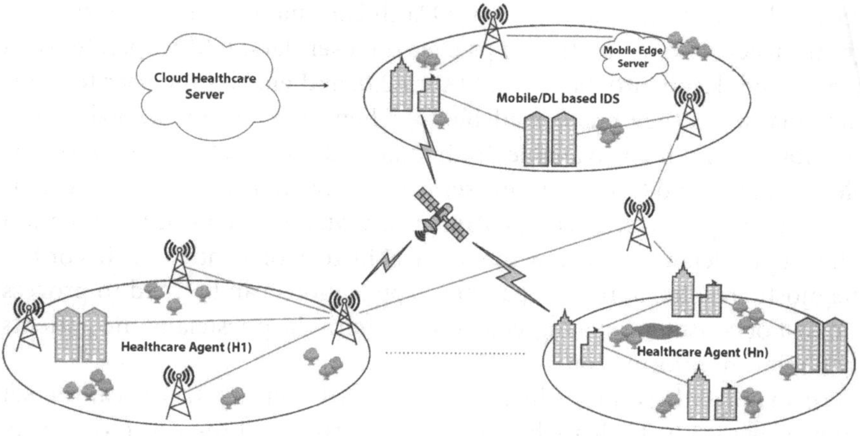

Figure 13.6 Workflow of intrusion detection system.

(b) Edge-Enabled Mobile Devices

The framework makes use of edge computing to overcome the drawbacks of conventional cloud computing, including network latency and connectivity problems. Security measures such as an IDS are implemented at the edge level, nearer the data source, because smart healthcare devices frequently lack the resources necessary to perform complex processing tasks. Real-time network [16] activity between healthcare cloud servers and healthcare agents is tracked and analyzed by these devices.

(c) Cloud-Enabled Health Server

The cloud server needs to store, scrutinize, and put the healthcare data to more stringent security checks. It also provides system administrators with a full picture of how the systems involved behave before, during, and after detection of cyberthreats, and revises security policies.

Such a framework will greatly reduce the level of interference with IoMT devices that have limited resources by deploying the IDS as a software-as-a-service (SaaS) application on the mobile-enabled edge devices. Although the edge-hosted IDS is used to deal with real-time threats, such devices transmit data. This SaaS model allows healthcare providers to successfully ensure security, and security updates occur in a timely manner with the latest threat intelligence and security patches. Also, it is possible to avail ourselves of services such as Shapley Additive Explanations (SHAP) within the cloud server to understand how decisions are made by the IDS, thus making them more transparent and responding to threats with more informed decisions.

(ii) Threat Model

Due to the sensitivity and Fourier transform of the data that they deal with, such as publicly available data to highly sensitive and officially classified as top-secret data, and depending on whether the data are structured or unstructured, smart healthcare environments are in a much more severe position regarding potential cyberattacks compared to the typical healthcare environments. The categories of the primary cyberthreats, which mobile healthcare systems should deal with, are the following:

(a) **Distributed denial-of-service attacks**: Such attacks send a high volume of junk traffic to the network to drain the system resources and deny legitimate users access to healthcare services. Such disruptions can cause major threats to the health of patients within a smart healthcare environment due to delays or impediment to the delivery of required medical services.

(b) **Access control breach**: Here, the hackers exploit vulnerabilities in authentication processes and gain access to healthcare systems illegally. This allows them to amplify privileges, drain sensitive data, or even install malware, which may affect the integrity of patient files and system functionalities.

(c) **Replay attacks**: Replay attacks help an attacker to deceive the users or clients by intercepting a legitimate interaction between the patient and the medical professional, and then reproducing the messages. This threatens the privacy of the data and can allow unauthorized access or unexpected behavior of the system.

(d) **Man-in-the-middle attacks**: These attacks involve eavesdropping on users and the healthcare application communications. The integrity and confidentiality of the system are put at risk in the event that the attacker alters communication or intercepts the information that is being discussed, which lies between people, and they are tricked.

(e) **Spoofing attacks**: These attacks give the attacker access to impersonate a bona fide device exploiting vulnerabilities of protocols, e.g., Address Resolution Protocol (ARP). With the help of this ruse, the attacker can take control over communications and perform malicious operations at discretion, at the expense of information security and the general security of the healthcare system.

13.5 Benefits and Applications of MCPSs in Health Optimization

MCPSs are indispensable to the enhancement of healthcare because they have numerous applications and demonstrated benefits. One of the primary advantages is proactive health management, which allows for spotting potential health-related risks early and implementing medical care on time. The patients get the responses instantaneously due to the continuous real-time monitoring, which significantly reduces the visits to the emergency department of the hospital many times in a row. As customizing care based on the specific information available to a specific patient will also lead to customized treatment plans, MCPSs will lead to more specific and more effective therapies. Moreover, the remote monitoring capabilities have facilitated cost efficiency due to the reduced need for physical visits and staying in a hospital. Through continuity of care and less dropout in the process, MCPSs have helped the older patients especially to lead better lives by ensuring enhanced better health outcomes and improved independence. MCPSs in health optimization have various applications such as remote patient monitoring, health data management, treatment recommendations, and patient tracking.

13.6 Conclusions and Future Directions

A combination of the emerging technologies and MCPSs helps massively in changing the healthcare services of the elderly. The results that AI will provide in terms of decision-making are going to improve the decision-making process and invoke trustworthiness and transparency. DTs guarantee the predictive diagnostics and individual prescription of the plan of treatment based on the active monitoring that will assist in managing chronic healthcare conditions in older adults. Human-centric healthcare ensures continuous, adaptive, and remote care offered through CPS Society 5.0. It would help to detect threats at their genesis, and safeguard patients' data integrity and privacy with the help of an IDS. Overall, the MCPSs in combination with the XAI, DTs, Society 5.0, and IDS concepts present a good path to the enhancement of healthcare and security in systems for an elderly population. This holistic view not only helps in meeting the medical requirements, but it also guarantees safety, confidence, and strength in mature health systems. With the necessity to focus on the vital technological enablers, the sphere of MCPSs is moving forward promptly so as to fulfill the demands of the contemporary world of medicine as fast as possible. To achieve it, i.e., preserve the confidentiality of patients, health-related confidential information must be sent and recorded securely. The most prominent concern is data privacy and security. Also, the interoperability plays an important role as it helps to make the attempts to integrate these different sensor platforms and systems with each other and make them exchange data successfully to promote it in different areas of healthcare. The issue of scalability also comes to the fore to enable MCPS solutions to serve new and larger populations without compromising on the performance and reliability of the healthcare systems as they expand. Current trends will bode well with edge AI and 6G integration that will reduce latency by making them smarter and using decentralized processing for healthcare monitoring. There is also a high demand for the use of DTs, which provide a real-time simulation of a patient's health condition to predict the course of events, optimize any specified course of treatment, and in general improve overall system performance in the context of personalized healthcare.

AI disclosure statement: The author confirms that no artificial intelligence (AI)-generated content or AI-assisted writing tools were used in the preparation of this chapter (including summaries and the abstract). All text, analysis, and conclusions are the result of the author's original work.

References

1. Liu, W., Zhao, F., Shankar, A., Maple, C., Peter, J.D., Kim, B.G., Explainable AI for Medical Image Analysis in Medical Cyber-Physical Systems: Enhancing Transparency and Trustworthiness of IoMT. *IEEE J. Biomed. Health Inform.*, 29, 2365, 2025.
2. Abououf, M., Singh, S., Mizouni, R., Otrok, H., Explainable AI for Event and Anomaly Detection and Classification in Healthcare Monitoring Systems. *IEEE Internet Things J.*, 11, 3446, 2024.
3. Nazar, M., Alam, M.M., Yafi, E., Su'ud, M.M., A Systematic Review of Human–Computer Interaction and Explainable Artificial Intelligence in Healthcare With Artificial Intelligence Techniques. *IEEE Access*, 9, 153316, 2021.
4. Chamola, V., Hassija, V., Sulthana, A.R., Ghosh, D., Dhingra, D., Sikdar, B., A Review of Trustworthy and Explainable Artificial Intelligence (XAI). *IEEE Access*, 11, 78994, 2023.

5. Roopa, M.S. and Venugopal, K.R., Digital Twins for Cyber-Physical Healthcare Systems: Architecture, Requirements, Systematic Analysis, and Future Prospects. *IEEE Access*, 13, 44963, 2025.
6. Sayin, B., Zoppi, T., Marchini, N., Khokhar, F.A., Passerini, A., Bringing Machine Learning Classifiers Into Critical Cyber-Physical Systems: A Matter of Design. *IEEE Access*, 13, 94858, 2025.
7. Fathima, A.S., Empowering Consumer Healthcare Through Sensor-Rich Devices Using Federated Learning for Secure Resource Recommendation. *IEEE Trans. Consum. Electron.*, 71, 1563, 2025.
8. Murmu, A., Kumar, P., Namasudra, S., Reddy, M.R., GAN-Based Federated Adversarial Learning for Enhancing Security Towards Consumer Digital Ecosystems. *IEEE Trans. Consum. Electron.*, 71, 1102, 2025.
9. Lv, Z. and Singh, A.K., Edge-Cloud-Based Wearable Computing for Automation Empowered Virtual Rehabilitation. *IEEE Trans. Autom. Sci. Eng.*, 21, 3896, 2024.
10. Huanca, F., Torres, M., Rodriguez-Fernandez, M., Núñez, F., A Cyber–Physical System for Real-Time Physiological Data Monitoring and Analysis *IEEE Internet Things J.*, 11, 28918, 2024.
11. Patil, R.V., Ambritta, N.P., Mahalle, P.N., Dey, N., Medical Cyber-Physical Systems in Society 5.0: Are We Ready?. *IEEE Technol. Soc. Mag.*, 3, 189, 2022.
12. Dutta, J. and Puthal, D., Advancing eHealth in Society 5.0: A Fuzzy Logic and Blockchain-Enhanced Framework for Integrating IoMT, Edge, and Cloud With AI. *IEEE Access*, 12, 195710, 2024.
13. Trivedi, J., Tahir, M., Isoaho, J., AI-Enhanced Threat Intelligence in Remote Patient Monitoring Systems: A Survey on Recent Advances, Challenges and Future Research Directions. *IEEE Access*, 13, 106465, 2025.
14. Wang and Wang, Z., Blockchain-Enhanced IoT Sensor Data Management for Engineering Monitoring. *IEEE Sens. J.*, 25, 7363, 2025.
15. Aljuhani, A., Alamri, A., Kumar, P., Jolfaei, A., An Intelligent and Explainable SaaS-Based Intrusion Detection System for Resource-Constrained IoMT. *IEEE Internet Things J.*, 11, 25454, 2024.
16. Chamola, V., Shall Peelam, M., Guizani, M., Niyato, D., Future of Connectivity: A Comprehensive Review of Innovations and Challenges in 7G Smart Networks. *IEEE Open J. Commun. Soc.*, 6, 3555, 2025.

Index

Also of Interest

Check out these other related titles from Scrivener Publishing

Self-Powered Cyber Physical Systems, Edited by Rathishchandra R. Gatti, Chandra Singh, Rajeev Agrawal, and Felcy Jyothi Serrao, ISBN: 9781119841883. This cutting-edge new volume provides a comprehensive exploration of emerging technologies and trends in energy management, self-powered devices, and cyber-physical systems, offering valuable insights into the future of autonomous systems and addressing the urgent need for energy-efficient solutions in a world that is increasingly data-driven and sensor-rich.

ENERGY STORAGE TECHNOLOGIES IN GRID MODERNIZATION, Edited by Sandeep Dhundhara, Yajvender Pal Verma, and Ashwani Kumar, ISBN: 9781119872115. Written and edited by a team of experts, this exciting new volume discusses the various types of energy storage technologies, the applications of energy storage systems, the performance improvement of modern power systems, their role in the real-time operation of power markets, and the operational issues of modern power systems, including renewable-based generating sources.

SMART GRIDS FOR SMART CITIES VOLUME 1, Edited by O.V. Gnana Swathika, K. Karthikeyan, and Sanjeevikumar Padmanaban, ISBN: 9781119872078. Written and edited by a team of experts in the field, this first volume in a two-volume set focuses on an interdisciplinary perspective on the financial, environmental, and other benefits of smart grid technologies and solutions for smart cities.

SMART GRIDS FOR SMART CITIES VOLUME 2: Real-Time Applications in Smart Cities, Edited by O.V. Gnana Swathika, K. Karthikeyan, and Sanjeevikumar Padmanaban, ISBN: 9781394215874. Written and edited by a team of experts in the field, this second volume in a two-volume set focuses on an interdisciplinary perspective on the financial, environmental, and other benefits of smart grid technologies and solutions for smart cities.

SMART GRIDS AND INTERNET OF THINGS, Edited by Sanjeevikumar Padmanaban, Jens Bo Holm-Nielsen, Rajesh Kumar Dhanaraj, Malathy Sathyamoorthy, and Balamurugan Balusamy, ISBN: 9781119812449. Written and edited by a team of international professionals, this groundbreaking new volume covers the latest technologies in automation, tracking, energy distribution and consumption of Internet of Things (IoT) devices with smart grids.

SMART GRIDS AND GREEN ENERGY SYSTEMS, Edited by A. Chitra, V. Indragandhi and W. Razia Sultana, ISBN: 9781119872030. Presenting the concepts and advances of smart grids within the context of "green" energy systems, this volume, written and edited by a global team of experts, goes into the practical applications that can be utilized across multiple disciplines and industries, for both the engineer and the student.

SMART GRIDS AND MICROGRIDS: Concepts and Applications, Edited by P. Prajof, S. Mohan Krishna, J. L. Febin Daya, Umashankar Subramaniam, and P. V. Brijesh, ISBN: 9781119760559. Written and edited by a team of experts in the field, this is the most comprehensive and up to date study of smart grids and microgrids for engineers, scientists, students, and other professionals.

MICROGRID TECHNOLOGIES, Edited by C. Sharmeela, P. Sivaraman, P. Sanjeevikumar, and Jens Bo Holm-Nielsen, ISBN: 9781119710790. Covering the concepts and fundamentals of microgrid technologies, this volume, written and edited by a global team of experts, also goes into the practical applications that can be utilized across multiple industries, for both the engineer and the student.

INTEGRATION OF RENEWABLE ENERGY SOURCES WITH SMART GRIDS, Edited by A. Mahaboob Subahani, M. Kathiresh and G. R. Kanagachidambaresan, ISBN: 9781119750420. Provides comprehensive coverage of renewable energy and its integration with smart grid technologies.

Encyclopedia of Renewable Energy, by James G. Speight, ISBN: 9781119363675. Written by a highly respected engineer and prolific author in the energy sector, this is the single most comprehensive, thorough, and up to date reference work on renewable energy.

Green Energy: Solar Energy, Photovoltaics, and Smart Cities, Edited by Suman Lata Tripathi and Sanjeevikumar Padmanaban, ISBN: 9781119760764. Covering the concepts and fundamentals of green energy, this volume, written and edited by a global team of experts, also goes into the practical applications that can be utilized across multiple industries, for both the engineer and the student.

Energy Storage, Edited by Umakanta Sahoo, ISBN: 9781119555513. Written and edited by a team of well-known and respected experts in the field, this new volume on energy storage presents the state-of-the-art developments and challenges in the field of renewable energy systems for sustainability and scalability for engineers, researchers, academicians, industry.

Energy Storage 2nd Edition, by Ralph Zito and Haleh Ardebili, ISBN: 9781119083597. A revision of the groundbreaking study of methods for storing energy on a massive scale to be used in wind, solar, and other renewable energy systems.

Hybrid Renewable Energy Systems, Edited by Umakanta Sahoo, ISBN: 9781119555575. Edited and written by some of the world's top experts in renewable energy, this is the most comprehensive and in-depth volume on hybrid renewable energy systems available, a must-have for any engineer, scientist, or student.

Progress in Solar Energy Technology and Applications, Edited by Umakanta Sahoo, ISBN: 9781119555605. This first volume in the new groundbreaking series, Advances in Renewable Energy, covers the latest concepts, trends, techniques, processes, and materials in solar energy, focusing on the state-of-the-art for the field and written by a group of world-renowned experts.

A Polygeneration Process Concept for Hybrid Solar and Biomass Power Plants: Simulation, Modeling, and Optimization, by Umakanta Sahoo, ISBN: 9781119536093. This is the most comprehensive and in-depth study of the theory and practical applications of a new and groundbreaking method for the energy industry to "go green" with renewable and alternative energy sources.

DESIGN AND DEVELOPMENT OF EFFICIENT ENERGY SYSTEMS, Edited by Suman Lata Tripathi, Dushyant Kumar Singh, Sanjeevikumar Padmanaban, and P. Raja, ISBN: 9781119761631. Covering the concepts and fundamentals of efficient energy systems, this volume, written and edited by a global team of experts, also goes into the practical applications that can be utilized across multiple industries, for both the engineer and the student.

INTELLIGENT RENEWABLE ENERGY SYSTEMS: Integrating Artificial Intelligence Techniques and Optimization Algorithms, Edited by Neeraj Priyadarshi, Akash Kumar Bhoi, Sanjeevikumar Padmanaban, S. Balamurugan, and Jens Bo Holm-Nielsen, ISBN: 9781119786276. This collection of papers on artificial intelligence and other methods for improving renewable energy systems, written by industry experts, is a reflection of the state of the art, a must-have for engineers, maintenance personnel, students, and anyone else wanting to stay abreast with current energy systems concepts and technology.

SMART CHARGING SOLUTIONS FOR HYBRID AND ELECTRIC VEHICLES, Edited by Sulabh Sachan, Sanjeevikumar Padmanaban, and Sanchari Deb, ISBN: 9781119768951. Written and edited by a team of experts in the field, this is the most comprehensive and up to date study of smart charging solutions for hybrid and electric vehicles for engineers, scientists, students, and other professionals.